Simulating Switching Converters with LTspice

SPEED UP YOUR POWER CONVERSION PROJECTS WITH READY-MADE TEMPLATES

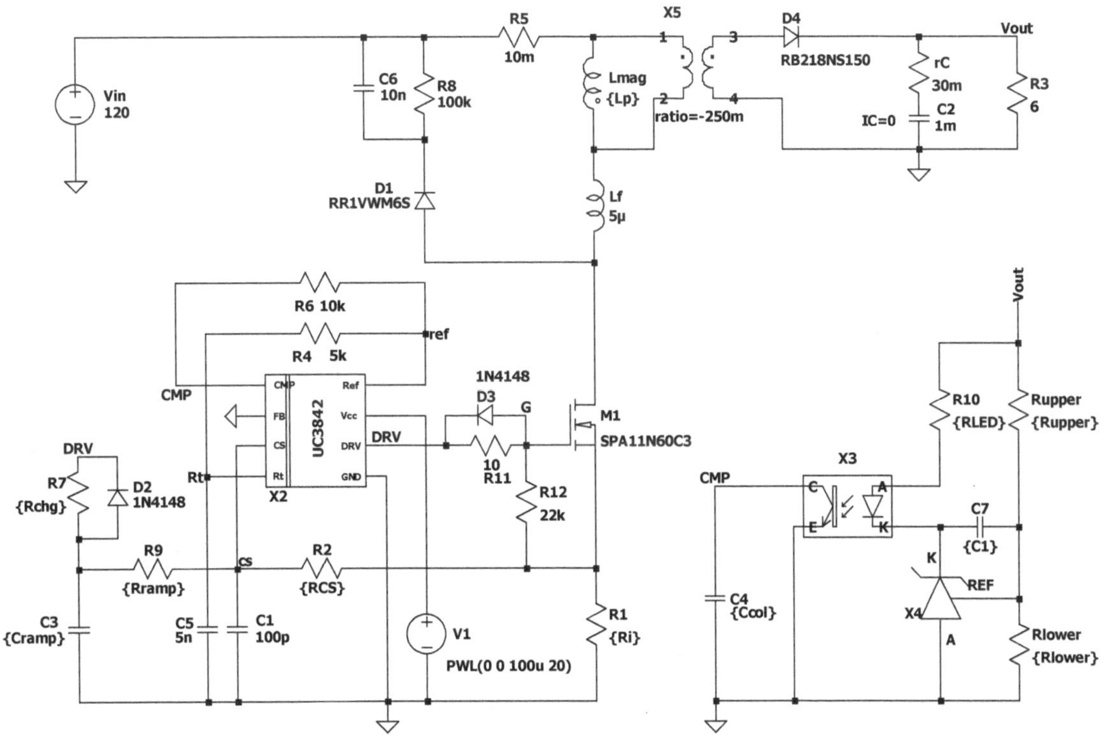

Christophe Basso

Simulating Switching Converters with LTspice

Other books by Christophe Basso:

An Intuitive Guide to Compensating Switching Power Supplies: Apply Christophe Basso's Recipes on Loop Control (2024)
The Fast Track to Determining Transfer Functions of Linear Circuits: The Student Guide (2023)
Transfer Functions of Switching Converters: Fast Analytical Techniques at Work with Small-Signal Analysis (2021)
Linear Circuit Transfer Functions: An Introduction to Fast Analytical Techniques (2016)
Switch-Mode Power Supplies, Second Edition: SPICE Simulations and Practical Designs (2014)
Designing Control Loops for Linear and Switching Power Supplies: A Tutorial Guide (2012)
Switch-Mode Power Supplies: Spice Simulations and Practical Designs (2008)
Switch-Mode Power Supply SPICE Cookbook (1996)

Print ISBN 978-1-960405-64-7
eBook ISBN 978-1-960405-65-4

Cover Design by Guy D. Corp, www.GrafixCorp.com

Faraday Press
1000 West Apache Trail—Suite 126
Apache Junction, AZ 85120 USA

Acknowledgments

I WOULD LIKE to warmly thank Alain Laprade (USA) who kindly reviewed this book for identifying typos but also for correcting my English. I know you also have tested all the LTspice examples, *très sympa* Alain! My friend Joël Turchi (France) was also part of the review team and made useful suggestions: *merci* Jojo. Finally, Eleazar Falco (Germany) thoroughly reviewed the manuscript as he kindly did for the previous title: *gracias mi amigo*.

Thank you to my sweet wife, Anne, who believed my previous title was the last one, but no, I still have plenty of ideas for publishing *mon Trésor*!

Merci also to Ken Coffman with Stairway Press, who spent time shaping my manuscript into the new book you hold: this is the fourth in the Faraday Press imprint and the 11th of a series started in 1996!

Preface

I STARTED SIMULATING switching power supplies back in 1989, during my eight-year period with the European Synchrotron Radiation Facility (ESRF) in Grenoble. At that time, the simulator of choice was PSpice from Microsim and a batch file – who remembers the .bat extension? – had been created between the OrCAD schematic capture program and the DOS-based simulation engine. Needless to say, it was often crashing, simulation was slow and I had to regularly tweak the netlist to make sure all was working in the end.

At that same period of time, I discovered the Intusoft ICAP/4 suite, programmed by Larry Meares, the founder of the company. Larry was well versed into power supply design and few people know he discovered one version of the PWM switch model at about the same time Dr. Vatché Vorpérian did, in 1986. Vatché did acknowledge that fact in a letter he wrote to the VPEC members 3 years later and also in his APEC 2017 plenary talk. The ICAP/4 suite had many excellent features – especially for manipulating waveforms – with a robust convergence algorithm, making it a serious contender to PSpice which became OrCAD's PSpice (now CADENCE's). Unfortunately, despite many numerous great additions over the years, the program could not keep up with the many players in that field, especially LTspice which was free and reliable.

I did not spend time with LTspice in the beginning, as my readers converted many of the models and circuits, originally written in different syntaxes. I have used the program time to time over the years, but did not really invest into it. Until a few years back, when I had to document 3-phase PFCs operations for training purposes. At that time, I found people claiming they had working circuits with nicely-documented waveforms in well-written blogs or published papers, but were they sharing files? Nope.

This situation triggered the need to modestly improve the situation. I worked for a year or so, developing 3-phase power factor correction circuits with various control methods (dq0, space vector modulation, hysteretic etc.) and all of these working circuits have been released in a ZIP file anybody can freely download. This is what I call sharing the knowledge and it is part of engineering excellence in my point of view. Finally, as I did with my free 150+ ready-made templates for SIMPLIS, I worked out a series of templates for LTspice that I am going to describe in this short manual.

I won't include too much of theory in this document considering the many books I have already published on the subject of power conversion. On the contrary, I will focus on practical examples you can immediately run and tweak according to your needs. Some of these examples are simple, some more complicated, so start at your own pace and always modify a working circuit via small steps – *petit à petit* – so that you can immediately roll back if it hangs and identify the offender.

This guide is for you, power supply designers, and, as usual, I will be happy to read comments you can send at *cbasso@orange.fr*. *Bonnes simulations*! Christophe Basso

About the Author

CHRISTOPHE BASSO WORKS as a business development manager for Future Electronics, in France. In his role, he provides technical assistance to customers developing power switching converters in Europe. Before this position, he was a Technical Fellow with *onsemi* in Toulouse, France. He led an application team dedicated to developing new offline PWM controller's specifications. Christophe has originated numerous integrated circuits among which the NCP120X series has set new standards for low standby power converters.

Christophe has released several books on power conversion and simulation. His last title was published with Stairway Press and is entitled *An Intuitive Guide to Compensating Switching Power Supplies*. This 170-page practical manual, contains a bit of theory but quickly goes into the design of control loops for switching converters. Using SIMPLIS as a simulation engine, it offers numerous design examples you can quickly reuse for your own projects. It can be seen as a very good complement to the present title.

Christophe has over 25 years of power supply industry experience. He holds 25 patents on power conversion and often participates in conferences and trade magazines including How2Power. Prior to joining *onsemi* in 1999, Christophe was an application engineer at Motorola Semiconductor in Toulouse. Before 1997, he worked for 8 years at the European Synchrotron Radiation Facility (ESRF) in Grenoble, France. He holds a *diplôme universitaire de technologie* from the university of Montpellier (France) and a MSEE from the Institut National Polytechnique de Toulouse (France). He is an IEEE Senior Member.

This is me, in the woods of French Catalonian mountains – near *Coll de la Llosa* – riding an old Cannondale Jekyll. My first mountain bike was a Marin Muirwoods, bought in 1989. Still working but my back can't afford hard tails anymore! Ok, to be honest, I moved to an electric mountain bike after I turned 60 ^_^

Table of Content

Table of Content

The Subcircuits

The Compensators

Introduction

WHEN I START simulating a switching converter, I always keep the schematic at the simplest possible level. Rather than assembling complex manufacturers models from scratch, I start with my own generic versions and verify that the circuit converges this way. When the simulation results suit my needs, I update the circuit with more precise or faster models, such as the ones supplied with LTspice or those developed by third-party manufacturers. If simulation fails at some point, I will know which subcircuit or model is guilty and I can fix it.

Why doing it this way? Because the many subtle interactions between SPICE components – the native like the *.model* or external such as a *.subckt* – can create convergence problems or have the circuit deliver wrong results. Running many of these subcircuits at once for a first attempt, complicates the analysis to identify the ones causing trouble.

This is the philosophy I have adopted for years and also applied in this manual. You will see it in action for op-amps but also for logic gates, transformers models and so on. The idea is to build the circuit yourself so a) you gain understanding on how it really works and b) you know how to explain and fix a convergence issue in case it happens. Once the entire converter operates as expected, then you can freely call more comprehensive models and progress with the simulation.

Ac analysis is an important part of the design and you know how much I care about compensation strategies. Closing the loop properly is key to building robust converters with a good and stable phase margin φ_m at the specified crossover frequency f_c. For that purpose, I was pleased to discover the frequency-response analyzer (FRA) symbol now offered by LTspice. It is fairly easy and convenient to use for extracting the compensated loop gain in switching circuits. It does not beat SIMPLIS in terms of execution speed and granularity (important for resonant circuits), but it truly represents a viable option for stabilizing the loop. I will show it in action in a few selected examples.

The LTspice ready-made templates you will discover in this book, can be used as design companions for testing your components values. For instance, should you need to design a flyback converter operated on a wide input range, you can immediately verify that the transformer you have designed (with a primary inductance and a turns ratio to start with), meets the power goal in various operating conditions: is the peak current in this transformer away from the specified saturation level (at high core temperature)? Can the converter pass 100% of the expected power level in the valley of the bulk capacitor with a 50-Hz rectified mains? Will the loop be stable for the first power-on sequence? What if the primary inductance varies, or if the leakage inductance is off by a few percent?

All these answers can be obtained fairly quickly by simulation and before ordering samples. My templates are easy to tweak, remembering that a generic controller is usually enough to test your values: no need for all the bells and whistles for a first pass. When the simplified circuit works, then add more complex components like MOSFETs, diodes etc. As a preliminary conclusion, a good simulation saves a tremendous amount of time but I am not saying anything new here!

The Op-Amp Generic Model

AMONG THE NEEDED subcircuits, the operational amplifier (op-amp) model represents an essential device. It has to reproduce the basic open-loop gain characteristic – a dc plateau, a low-frequency pole – and have clamped outputs to avoid saturating the downstream circuitry. A classical voltage-controlled voltage E-source can sometimes do the job (in simple ac sweeps), but it is complicated to smoothly clamp the output when the circuit rails up or down, at startup for instance. Should you want to implement maximum source or sink characteristics with the *E*-source, it becomes even trickier.

For years, I have adopted a simple voltage-controlled current *G*-source feeding a *RC* network (for the gain and the pole), followed by a pair of diodes clamping the output at the levels I want. Simple, efficient and always converging. If, later on, I want to see the effects of slew-rate, offsets and so on, then I can complexify this simple model or resort to a true macro model available in LTspice.

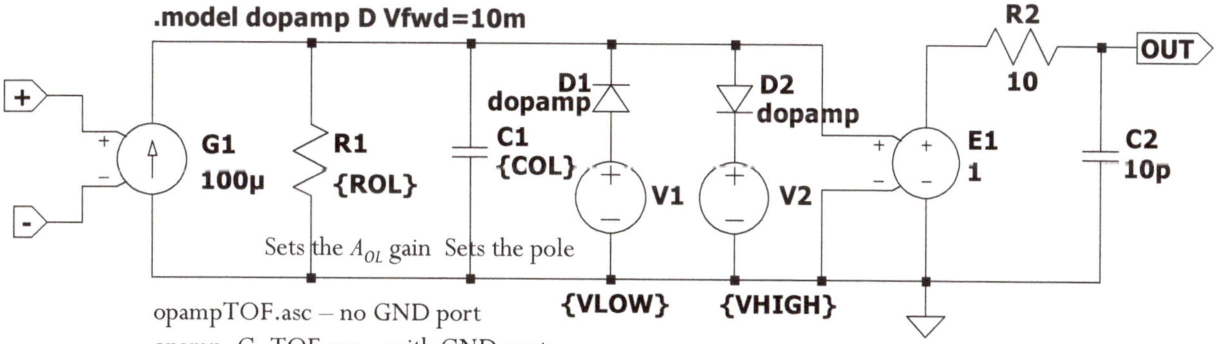

opampTOF.asc – no GND port
opamp_G_TOF.asc – with GND port

The dc open-loop gain, often designated as A_{OL}, is obtained by determining resistance R_1 value. Once you have that value, the low-frequency pole is determined by C_1. Select the maximum output voltages, e.g. 5 V and 100 mV or ±15 V, and you are all set! The GND port in the *opamp_G_TOF.asc* model, allows you to insert an op-amp referenced to any node you want.

```
*
* Choose op amp characteristics *
*
.param AOL=90 ; open-loop gain in dB *
.param POLE=30 ; low-frequency pole *
.param VHIGH=5 ; upper output level *
.param VLOW=100m ; lower output level *
*
* Do not edit these lines *
*
.param GAIN={10**(AOL/20)}
.param COL={1/(6.28*(GAIN/100u)*POLE)}
.param ROL={GAIN/100u}
*
```

This is an automated macro

You can place this type of automated sheet via the **.t** button found in the ribbo **t .t** All SPICE directives are captured using a text zone created with the **.t** button. The t button is for a normal text zone.

To visualize the computed results, insert in the macro, a *.options logparams* statement. All results will appear in the *.log file after the simulation has run. You can invoke the file by pressing <CTRL> + L.

Testing the Op-Amp Model

AFTER BUILDING the subcircuit, it is important to thoroughly test it for validation purpose. For a simple op-amp, you can verify that its open-loop response matches what is stated in the datasheet, for instance. However, one very important parameter to account for, is the dc operating point: before considering the ac results, you must be sure the op-amp works in its linear zone during the ac analysis, e.g. $V_{out}\approx 3$ V for a 0-5 V single supply, for example. If V_{out} rails up or down, then most likely the ac response will be wrong.

How to fix the dc bias with such a high open-loop gain? If you try to bias the input for a 3-V output, you may have to tweak the dc source at the µV level, something you want to avoid. Use the below auto-bias circuit to fix V_{out} to any value before the ac modulation.

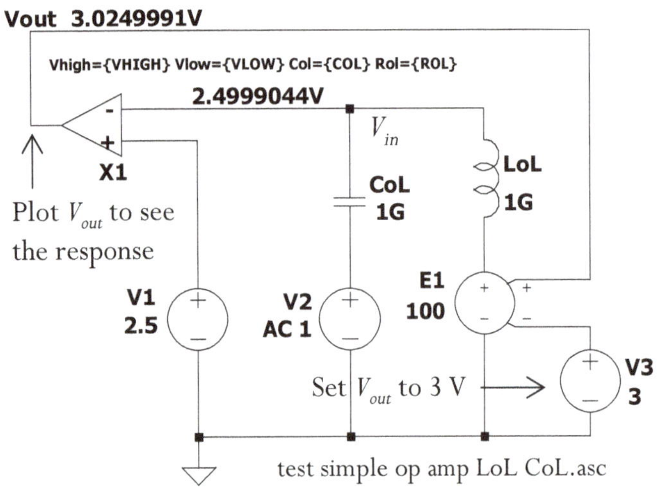

test simple op amp LoL CoL.asc

```
* Choose op amp characteristics *
*
.param AOL=90 ; open-loop gain in dB *
.param POLE=30 ; low-frequency pole *
.param VHIGH=5 ; upper output level *
.param VLOW=100m ; lower output level *
*
* Do not edit these lines *
*
.param GAIN={10**(AOL/20)}
.param COL={1/(6.28*(GAIN/100u)*POLE
.param ROL={GAIN/100u}
*
.options logparams
*
.ac dec 100 1 1Meg
```

Once the simulation is done, simply plot V_{out} and you have the open-loop response that I called G, showing a dc gain of 90 dB, with a low-frequency pole, as expected.

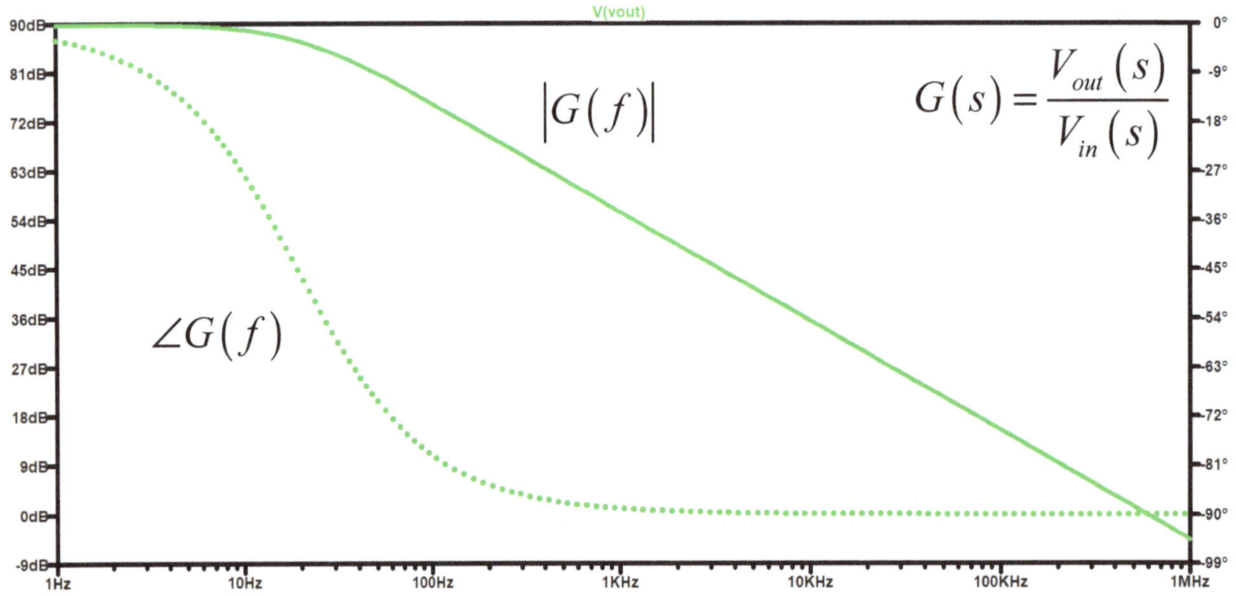

Setting the Bias Point

IN THIS EXAMPLE, the *LoL/CoL* network is a classic in this type of analysis. Before the ac simulation starts, LTspice will compute a dc bias point by opening all capacitors and shorting all inductors. Doing so, the loop is temporarily closed in dc via the low-gain voltage-controlled voltage source E_1 and the op-amp output is roughly set to around 3 V in this example, away from ground and the 5-V upper limit set by *VHIGH*. When the ac sweep runs, the *LC* network acts like a filter with an extremely low cutoff frequency, opening the loop in ac. As you can see, with a 90 dB of open-loop gain A_{OL}, the voltage at the inverting pin is adjusted at the µV level to meet the 3-V output voltage.

Having the ac source inserted in series with the output of E_1, offers an alternate way to plot the open-loop gain. You would simply need to display V_{out}/V_{in} in this mode.

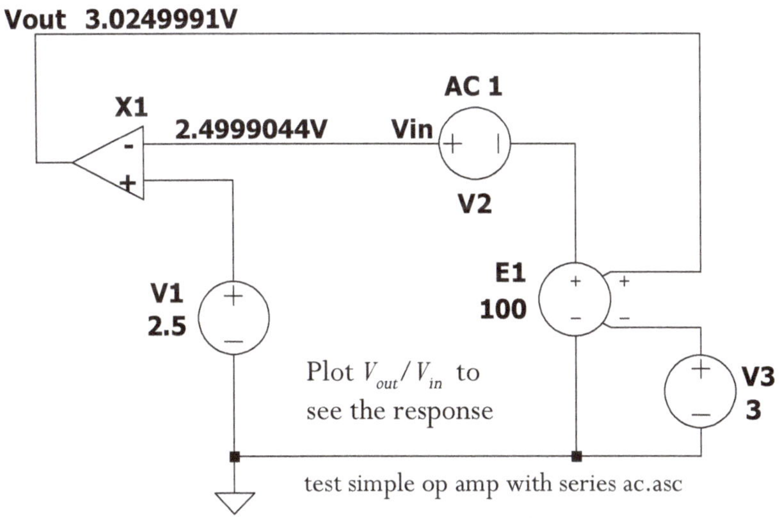

test simple op amp with series ac.asc

These are the values computed by the macro and located in the .LOG file.

```
Effective Parameter Values:
AOL = 9.000000000000000e+01
COL = 1.678491327053280e-11
GAIN = 3.162277660168379e+04
POLE = 3.000000000000000e+01
ROL = 3.162277660168380e+08
VHIGH = 5.000000000000000e+00
VLOW = 1.000000000000000e-01
x1:Col = 1.678491327053280e-11
x1:Rol = 3.162277660168380e+08
x1:Vhigh = 5.000000000000000e+00
x1:Vlow = 1.000000000000000e-01
```

There is another interesting approach, detailed by *a concerned citizen*, contributor on StackExchange. It uses an undocumented feature where you can have different resistances values for dc and ac analyses. In the below picture, the loop is closed in dc via the two resistances, while in ac, the new values impose a very high ratio, clamped by A_{OL}:

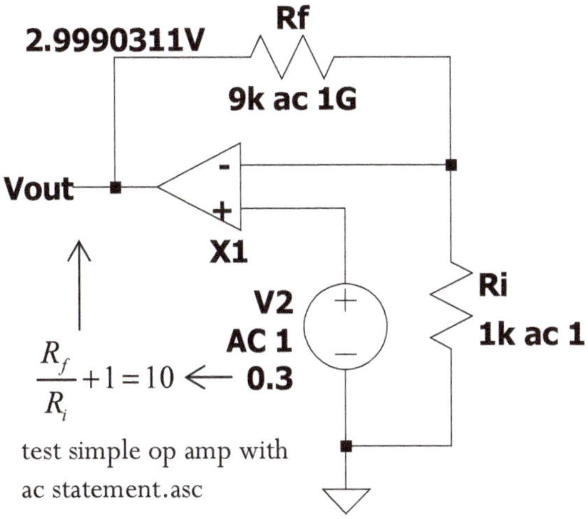

test simple op amp with ac statement.asc

The ac keyword followed by the resistance value in ac is simply added in the regular value field or in the *Value2* field (CTRL + right-click on the resistor).

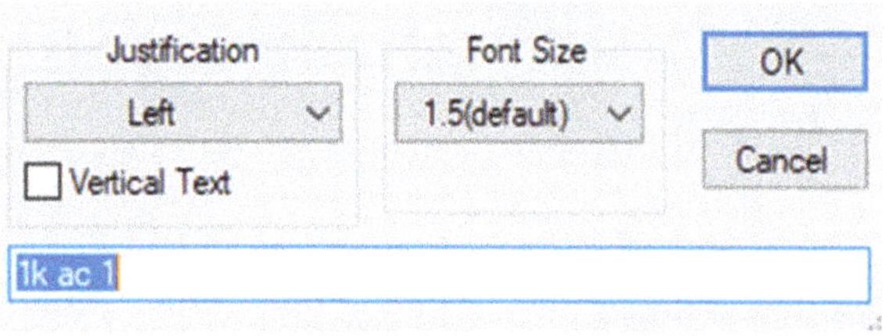

A More Comprehensive Version

THE SIMPLE model I have shown, suits most of the needs to test a regulation loop in normal conditions. Should you now want to load the op-amp and sink or source currents, the previous model is too simple to visualize loading effects. The below circuit includes the capability to adjust the maximum current the op-amp can deliver (source) or absorb (sink), before the output stage gives up and rails up or down. The model includes these options and brings a ground point for an easier 0-V connection in floating cases:

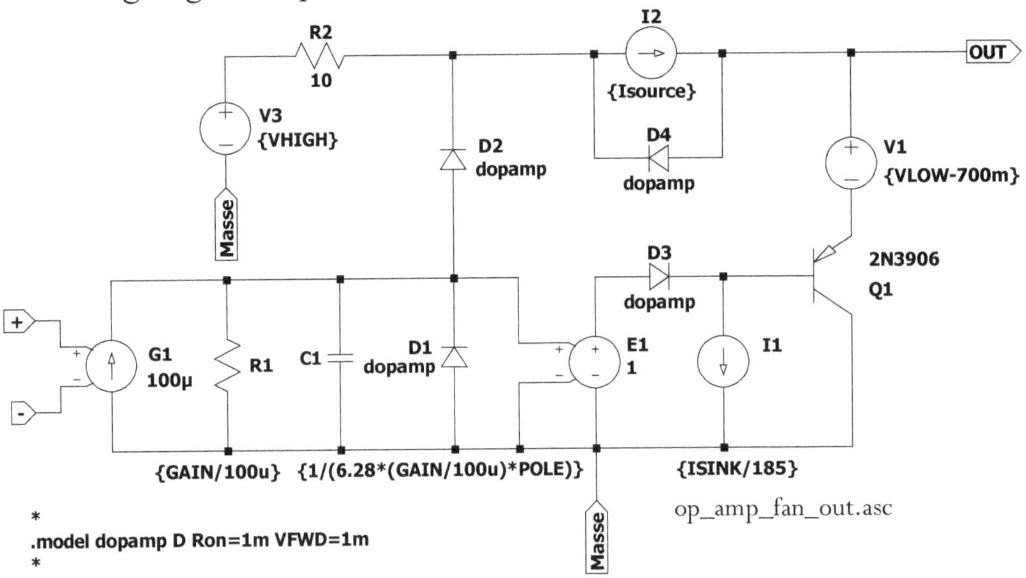

op_amp_fan_out.asc

With this model, by adding a second pole after the 0-dB transition (around 1 MHz) and including a 2.5-V voltage reference, you have a perfect candidate for the internal op-amp of the UC384x controllers. In this figure, the output cannot source and sink more than 500 µA and 6 mA of current, respectively.

op_amp_384x.asc

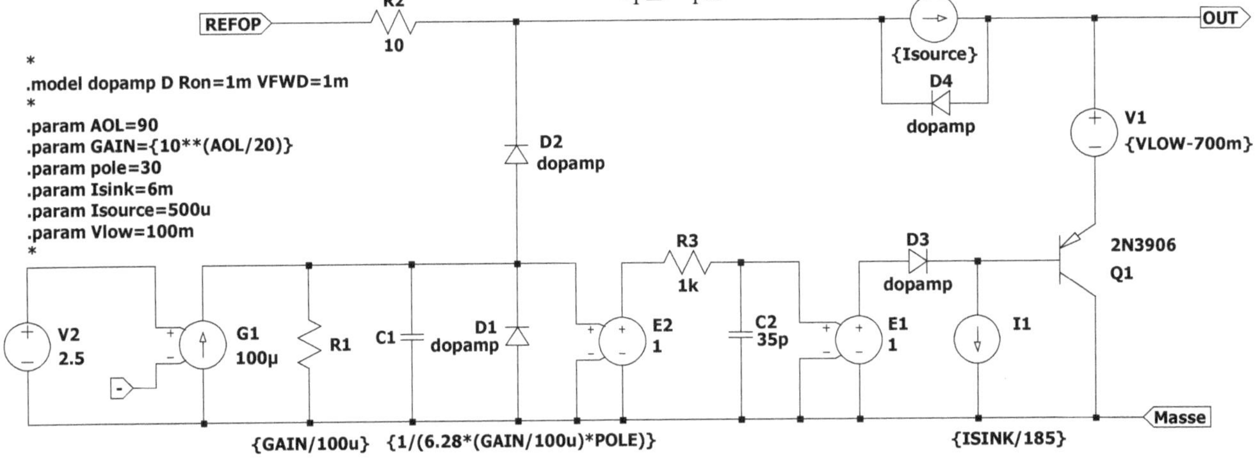

This model will be included in the UC384x subcircuits that I describe later in these pages. The 500-µA source current is perfect for disabling the internal op-amp and connect the collector of an optocoupler directly at its output, itself pulled-up to V_{ref} via an external resistance.

Testing the Model

BESIDE THE 2-pole frequency response, the new model features current sink and source limits that we must confirm before validating the subcircuit. The test circuits are quite simple and involve increasing the absorbed/injected currents on the op-amp output while set at high and low output levels, respectively:

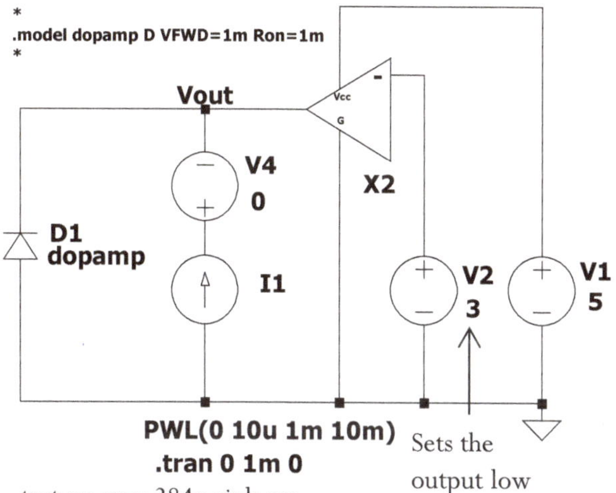

test op amp 384x sink.asc

The left-side circuit injects or sinks current into the op-amp (sink limit). The right-side version draws current from the op-amp (source limit).

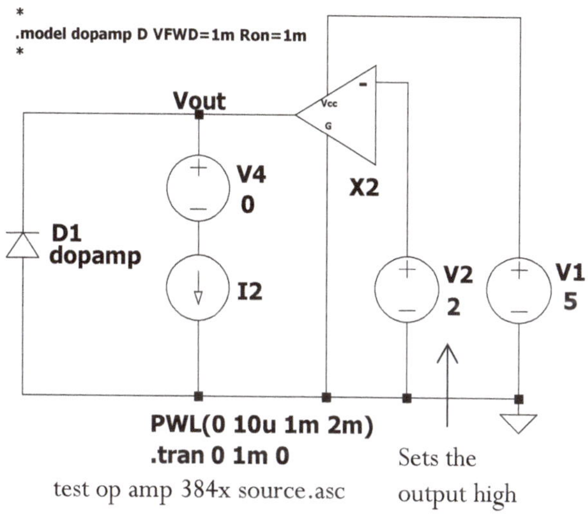

test op amp 384x source.asc

In the below graph, the output is in low state and tries to keep low while the current source injects current up to 6 mA. Beyond this value, the op-amp gives up and the output voltage rails up:

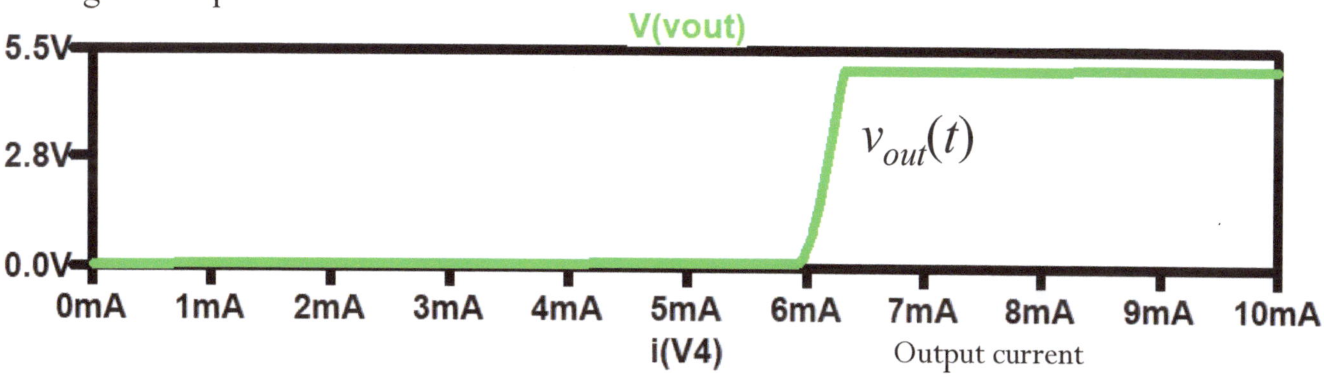

This graph shows the opposite situation where the op-amp rails up but a load draws current. When the current exceeds 500 µA, the output collapses to ground.

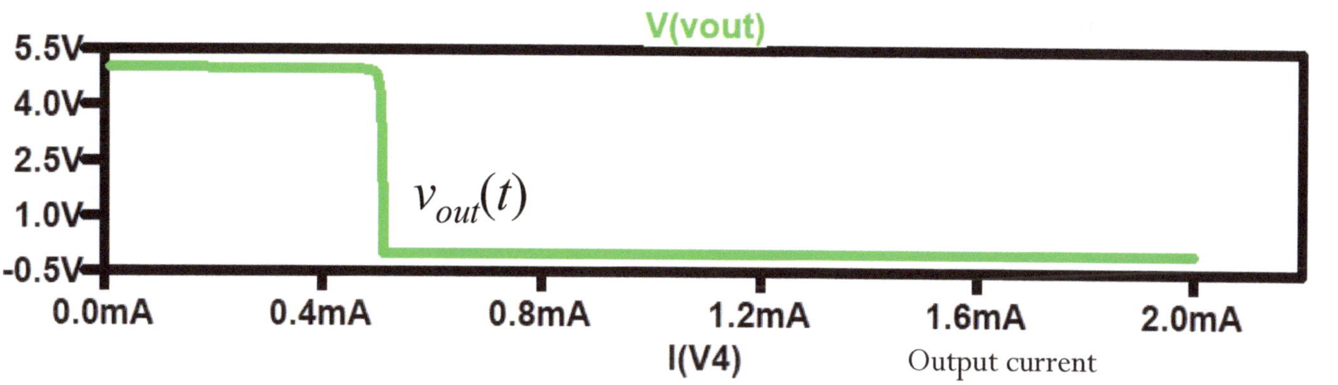

A Simple Transformer

I OFTEN SEE designers using the coupling coefficient *k* in their SPICE simulations for building a transformer with 2 or more windings. If this approach is perfectly valid, I prefer resorting to a more generic model [1] in which I can clearly see the leakage term and the transformer turns ratio. That way, I can adjust them independently, without recomputing the *k* value every time I make a change. Besides, I can plot the magnetizing current in a power transformer, independently from the primary current. It is interesting when you want to check a possible volt-seconds imbalance at start-up or steady-state in some specific structures like in an LLC converter, for instance. Finally, this transformer model works in dc, meaning I can use it with averaged models for ac analysis:

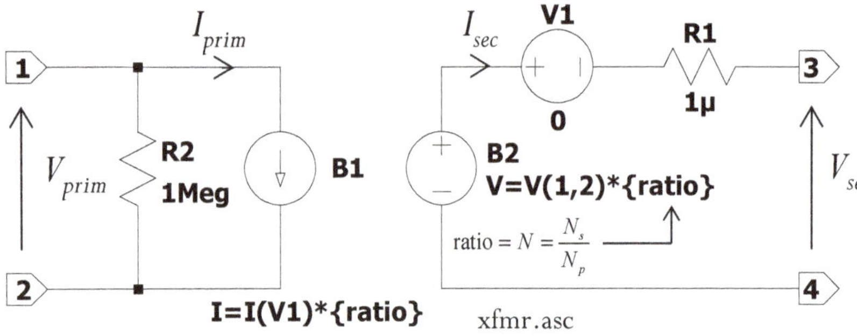

This simple circuit models a transformer with a turns ratio of 1:*N*, implying $V_{sec}=NV_{prim}$ and $I_{prim}=NI_{sec}$. If you have $N_p=10$, $N_s=2$, then $N=0.2$ or 200m. Of course, the model works in either directions. *N* can be negative, e.g. for a flyback.

Now, we need to find the correspondence between the coupling coefficient with the equivalent leakage term and turns ratio for this equivalent circuit.

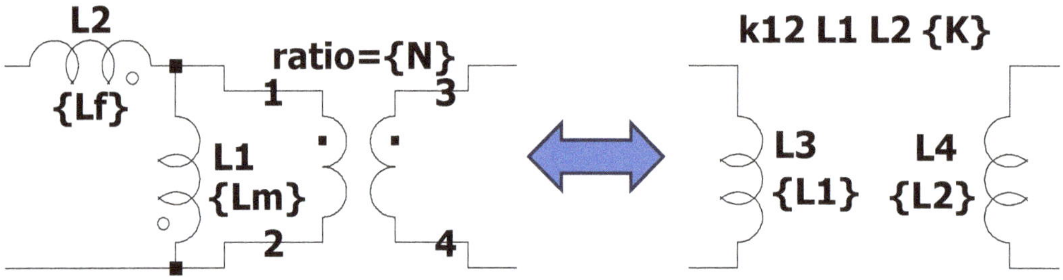

Assume you know the primary and leakage inductances with the turns ratio *N*, what is the coupling coefficient? The other way around, you have a coupling factor of 0.98, what are the transformer elements?

You know L_m, L_f and *N*:

$L_m := 600\mu H \quad l_f := 6\mu H \quad N_1 := 0.25$

$$k_1 := \frac{\sqrt{L_m \cdot (l_f + L_m)}}{l_f + L_m} = 0.995$$

$$L_1 := \frac{l_f}{1 - k_1^2} = 606\,\mu H$$

$$L_2 := L_1 \cdot N_1^2 \cdot k_1^2 = 37.5\,\mu H$$

With this type of transformer model, you could easily add saturation effects to the magnetizing inductance as described in [1], add inter-winding capacitances etc.

You know *k*, L_1 and L_2:

$L_1 := 606\mu H \quad k_1 := 0.995$

$L_2 := 37.5\mu H$

$$L_f := L_1 \cdot \left(1 - k_1^2\right) = 6.04485\,\mu H$$

$$L_m := L_1 \cdot k_1^2 = 599.95515\,\mu H$$

$$N_1 := \frac{1}{k_1} \cdot \sqrt{\frac{L_2}{L_1}} = 0.25001$$

Testing the Dual Winding Version

WE CAN AUTOMATE calculations through a simple macro to test our transformer model when fed by a sinusoidal stimulus.

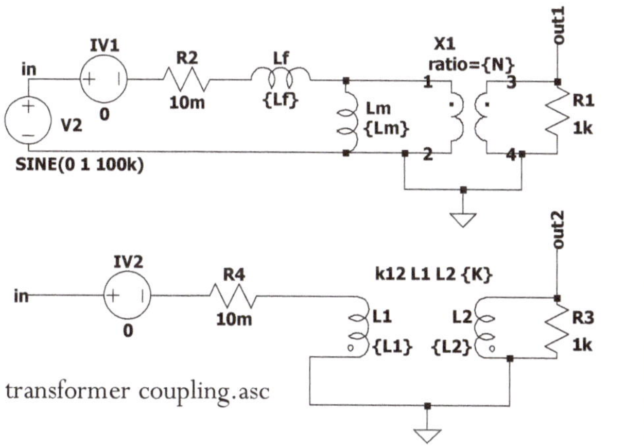

transformer coupling.asc

```
* parameters
*
.param k=0.995 ; coupling coefficient
.param L1=606u
.param L2=37.5u
.param Lf={L1*(1-k**2)}
.param Lm={L1*k**2}
.param N={(1/k)*sqrt(L2/L1)}
*
.options logparams
*
.tran 0 100u 0 100n
```

Save computed values to the .log file

This is one way to display the calculated values

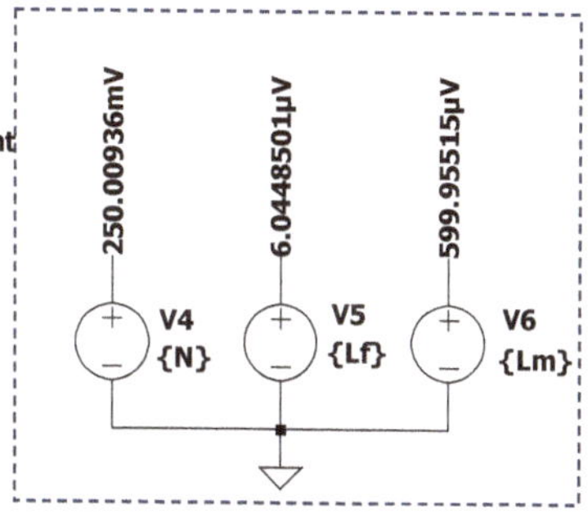

Dc values obtained after a .OP simulation.

The results obtained by the calculations can be displayed the "old" way through dc biases or via the *logparams* statement and the corresponding .log file. As shown below, the voltage/current amplitudes are rigorously identical between the two approaches:

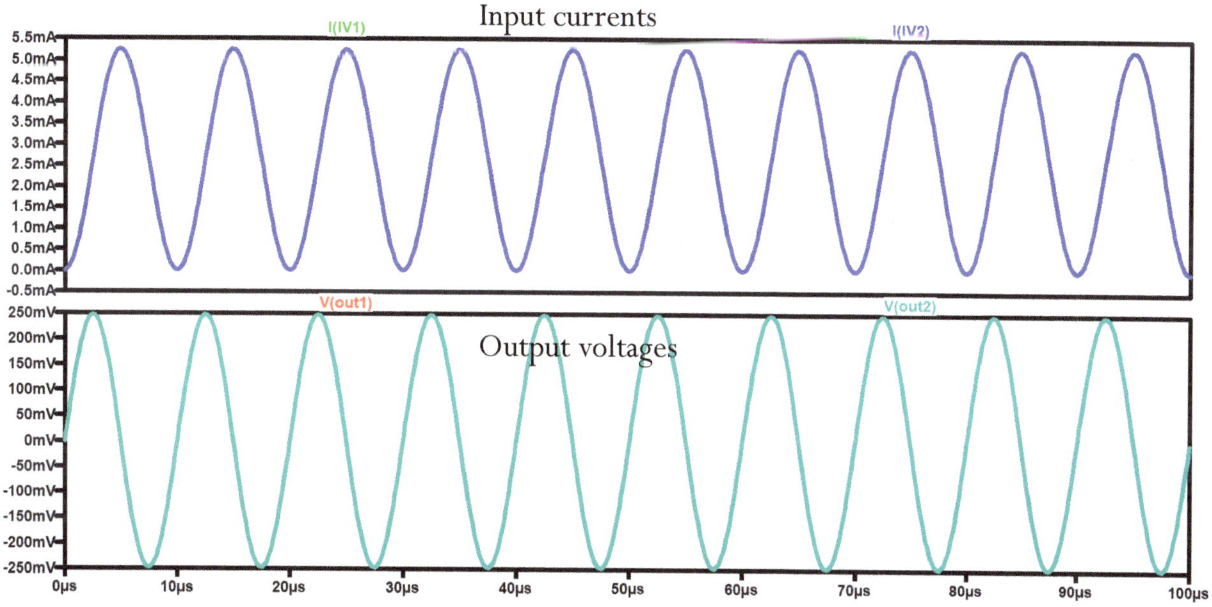

You can use this transformer model in a variety of applications, flyback, forward or as a current-sense transformer in which you can check the proper demagnetization process.

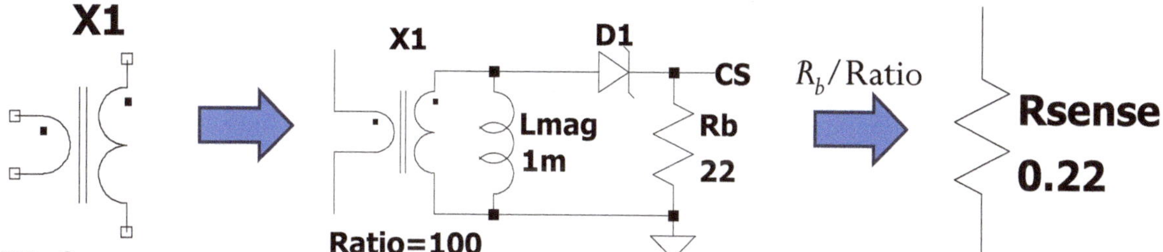

Multi-Winding Transformers

THE SITUATION DOES not really complicate with a 3-winding transformer, as shown below. It can easily be extended to more windings by adding the reflected currents on the primary side and accounting for the new turns ratios. They are denoted by *ratio12* and *ratio13* in this schematic diagram:

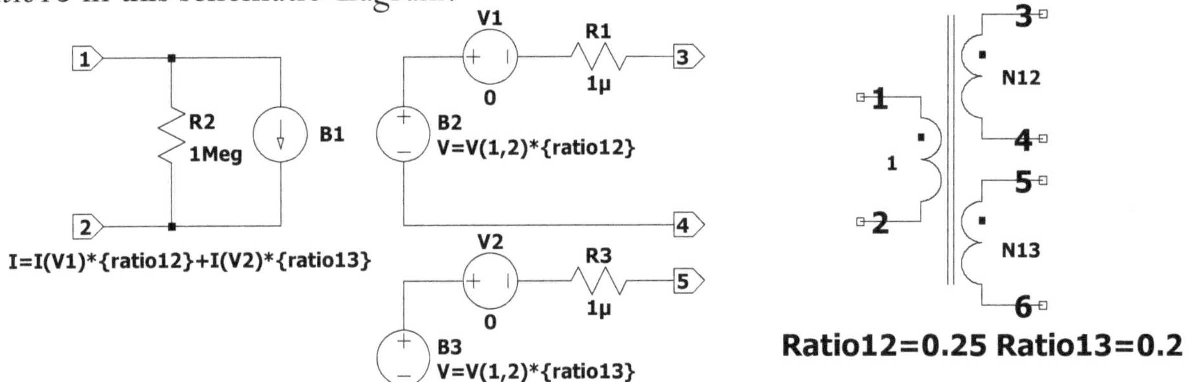

$1{:}N_{12} = 1{:}0.25$
$1{:}N_{13} = 1{:}0.2$

A center-tap version in the secondary is often needed in LLC converters, for instance, if two diodes are used. Very similar construction to the one above:

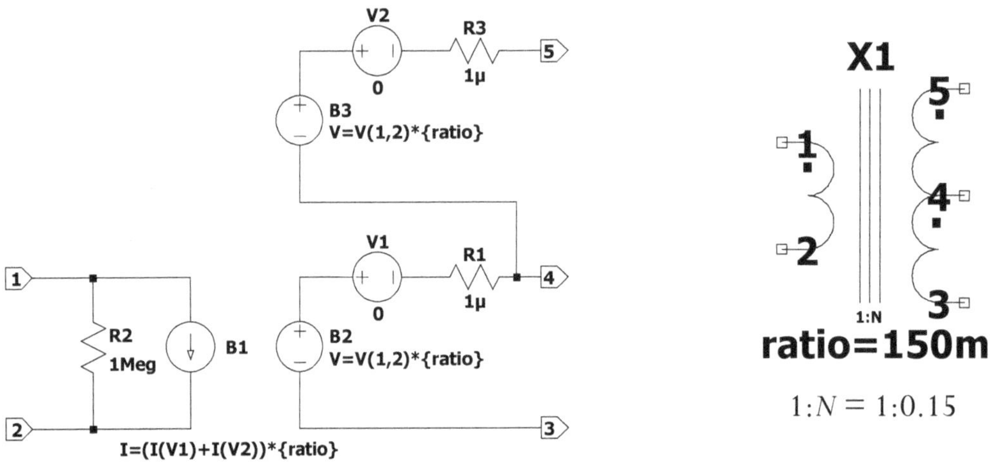

$1{:}N = 1{:}0.15$

All these transformers can conveniently be inserted in flyback or forward converters, LLCs and so on.

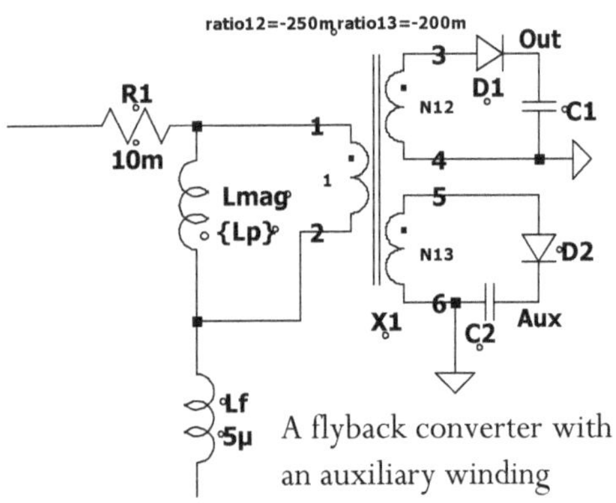

A flyback converter with an auxiliary winding

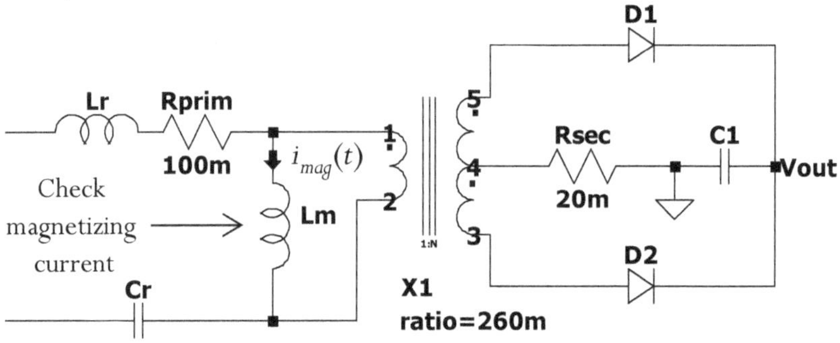

$1{:}N = 1{:}0.26$

An LLC converter with center-tap rectification

A Push-Pull Transformer

THE PUSH-PULL TRANSFORMER requires a bit more care to assemble it. The circuit below associates various sources to mimic the behavior of this type of transformer:

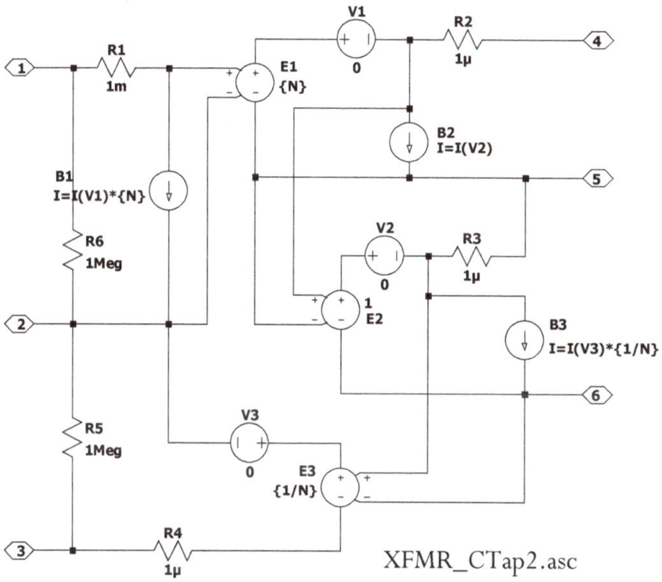

XFMR_CTap2.asc

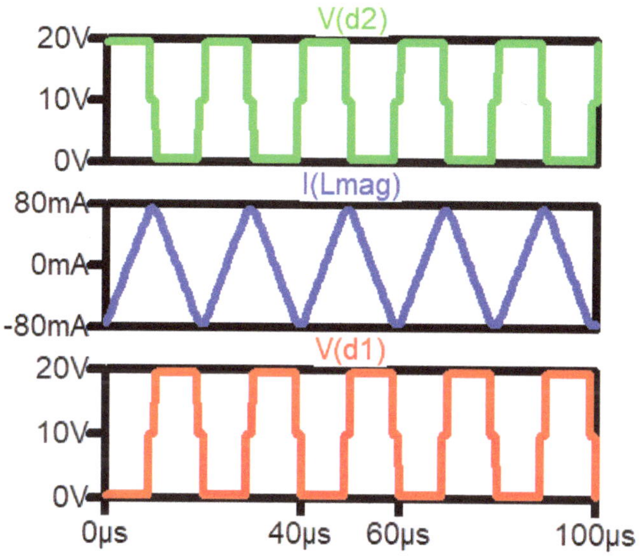

In this model, the magnetizing inductance is added between the tap and one of the primary-side pins, 1 or 3. As shown in the above right curves, the magnetizing current is well symmetrical and confirms the transformer is driven in quadrants I and III. We will come back to this example when tackling the push-pull example in voltage- and current-mode control.

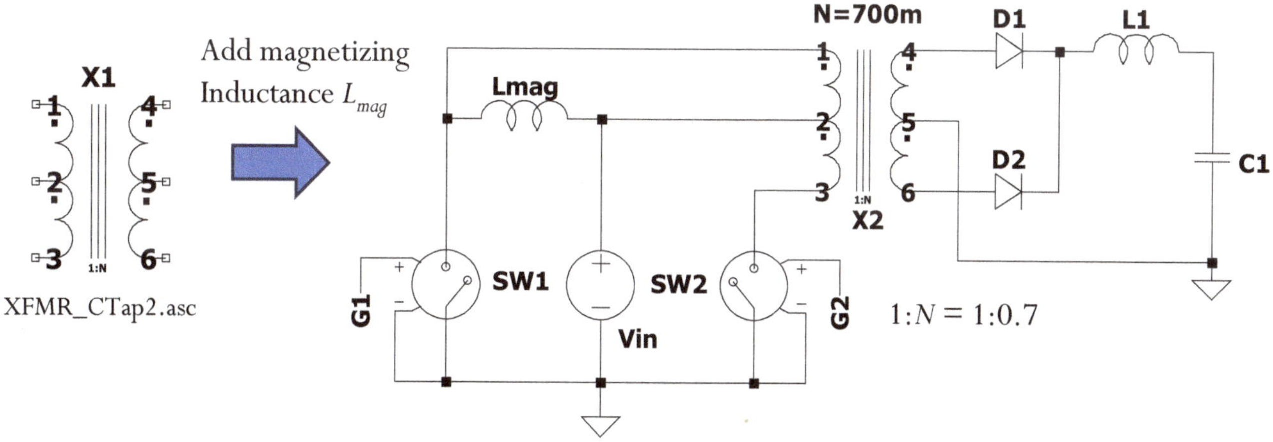

The electrical circuit above shows the simplified connection of the transformer with two ideal switches. The model should be supplemented with leakage inductances as they can potentially induce large voltage transients across the switches and represent a difficult obstacle in high-power converters. There are plenty of integrated controllers that are available for building this type of power supply (LM5033, LTC3721-1 or the older SG3525A), but I will resort to a generic model in the dedicated section, later in this manual.

A Simple SR Latch

THIS TYPE OF SUBCIRCUIT is found in various places when building a PWM controller. I have used a simple association of analog behavioral sources – the B types – in which I have coded OR gates with the | symbol. An extra test is added to check that outputs *Q* and Q_b are never high at the same time. The circuit works fluidly and I have tested it in many different configurations; it is really a cornerstone for all my models.

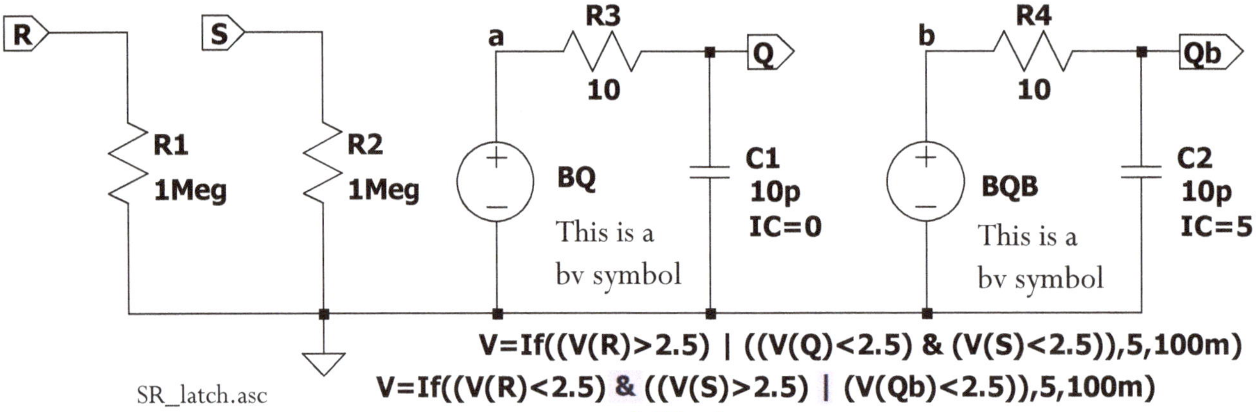

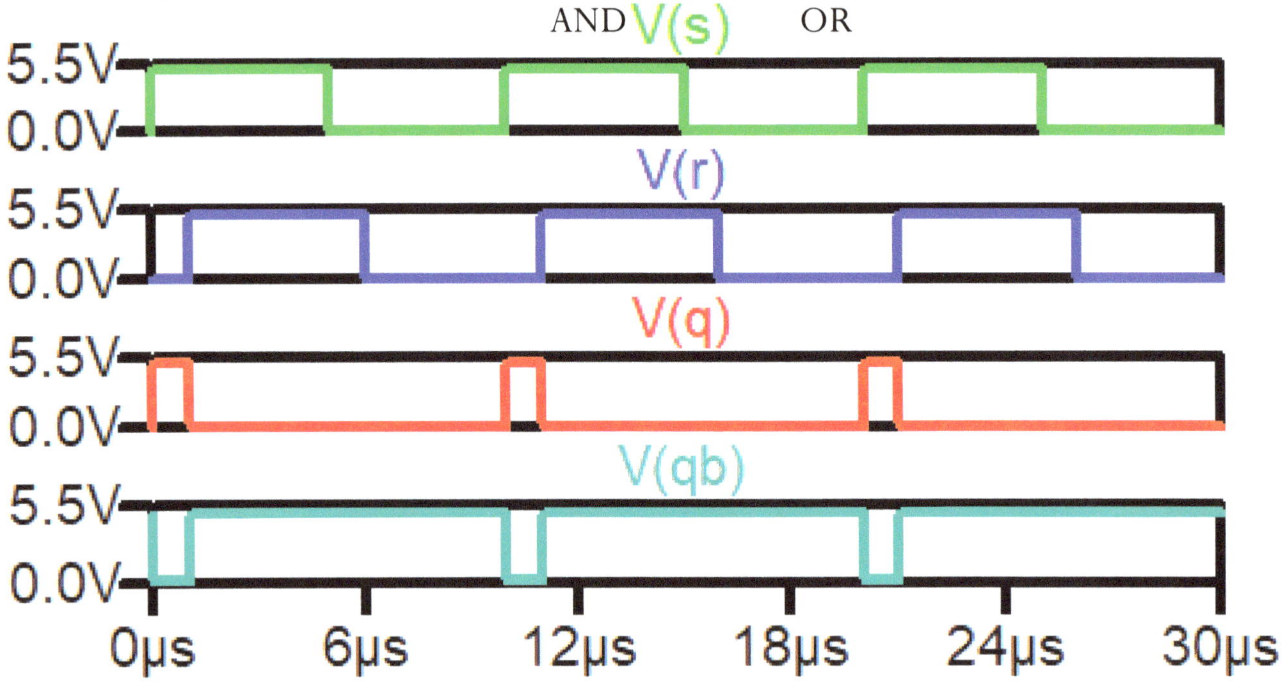

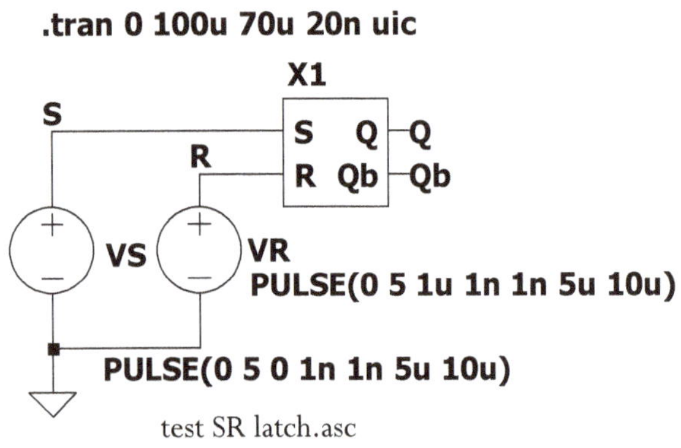

In this test, the latch is set high by the S node, and later on, the reset occurs with R going high, while S is still active: the output *Q* is correctly set low in this forbidden state and Q_b works in the expected complementary way which is good!

Check the *use initial conditions* box on the simulation panel - UIC

A D-Flip-Flop

THE D-FLIP-FLOP builds on the SR latch model, to which I have added a bunch of gates, as described in various sites. There are plenty of ways to build this type of logic circuit and this one gave good results in the simulations. I am sure the native LTspice model simulates faster but I got this one running reliably.

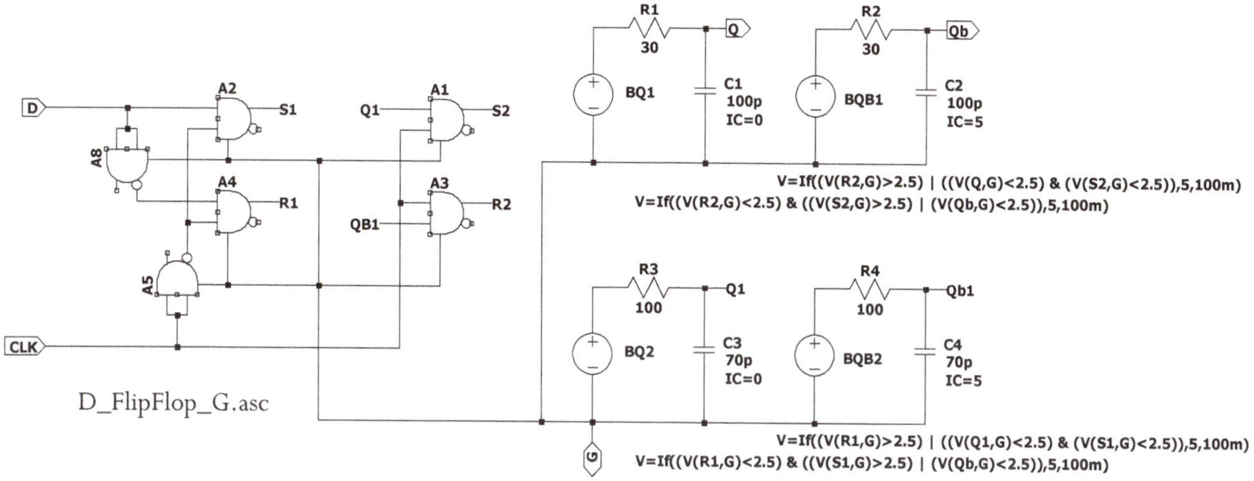

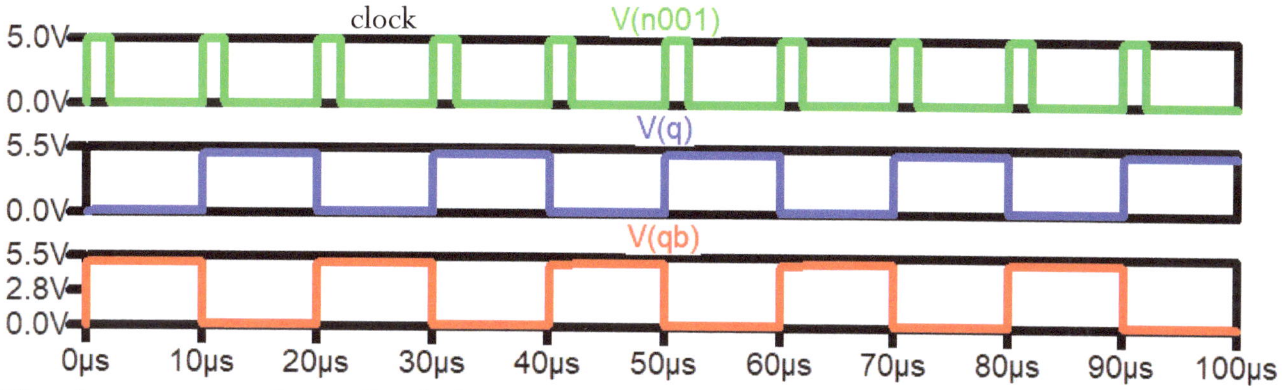

The device is wired in a divide-by-2 circuit and the above waveforms confirm it does the job as expected. A version with a reset pin has also been developed, by simply adding an OR gate to the R2 node. The rest of the circuit remains unchanged.

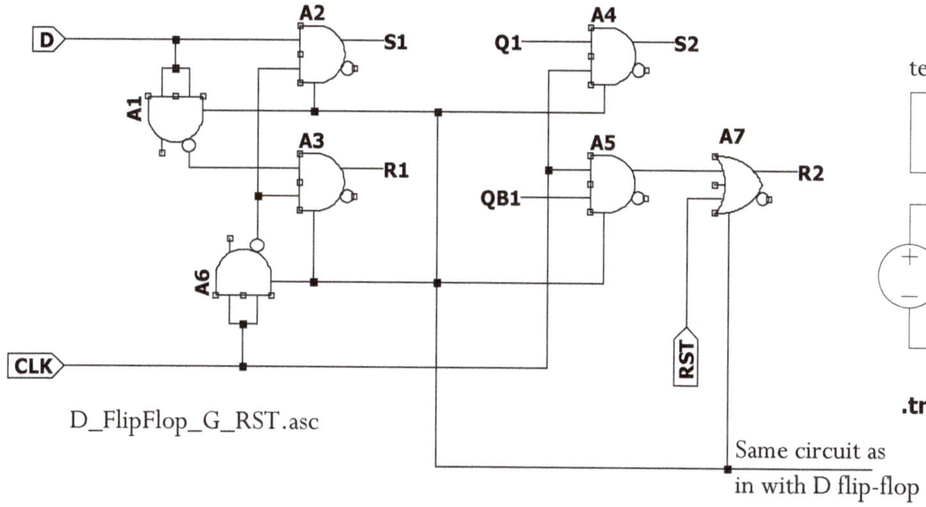

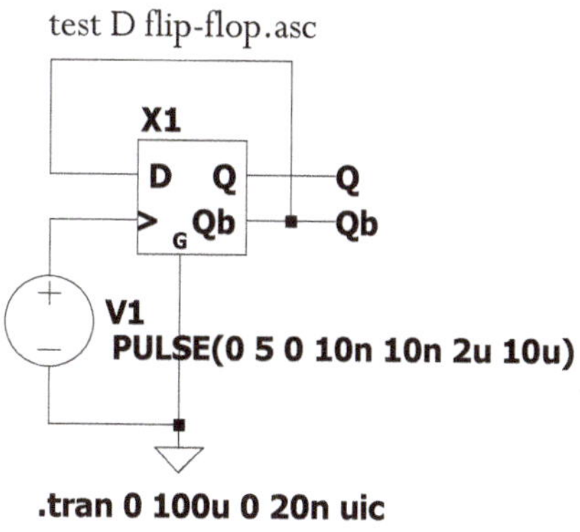

A Basic Comparator

THE COMPARATOR is a key subcircuit in switching converters. Two simple in-line equations will describe the function by adding some programmed hysteresis. A hysteresis is important to improve the commutation in case of a signal either noisy or having a low voltage slope. A ground pin has been added in case the comparator is not referenced to ground. The subcircuit includes a *RC* network to slow down the output transition.

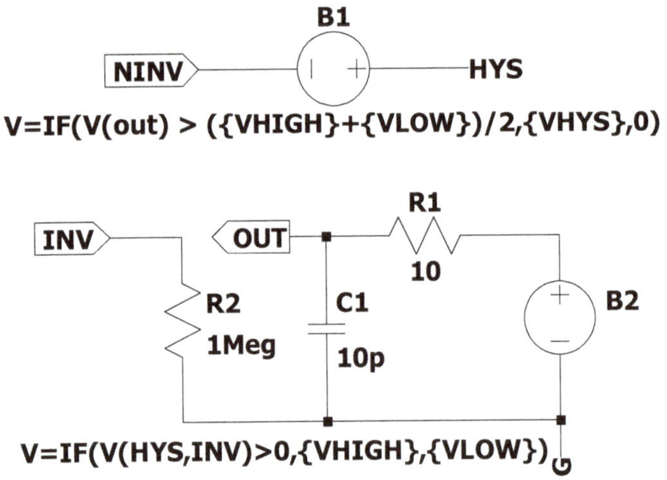

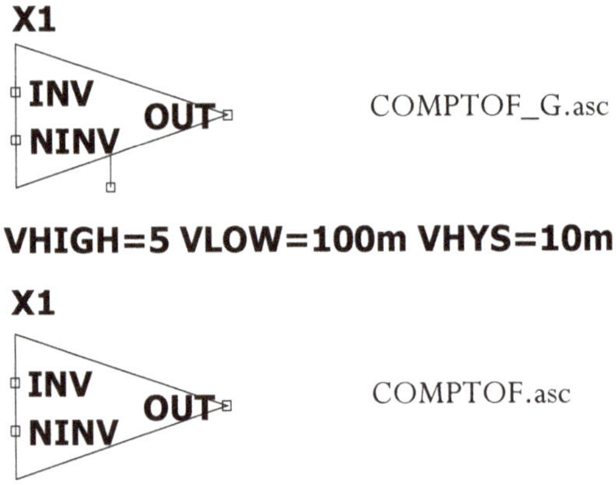

This is the version without ground

The test circuit is quite simple: a triangular waveform drives the non-inverting pin while the inverting pin is biased at 1 V. The hysteresis is set at 100 mV in this example:

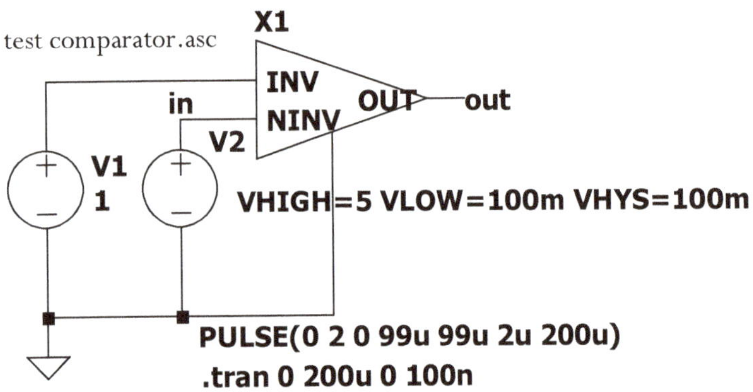

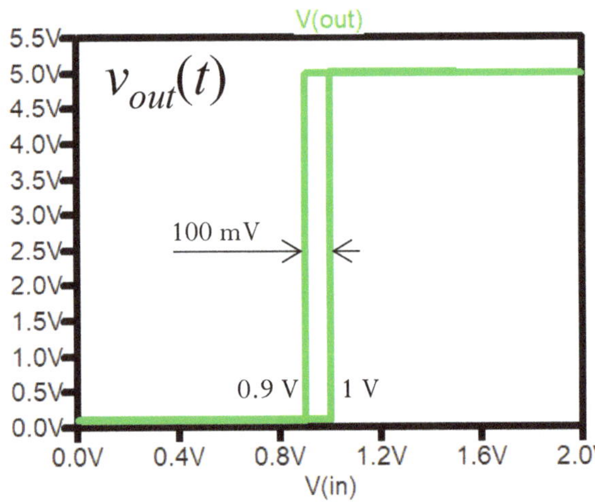

It is also possible to insert a propagation time by routing the comparator output through a simple buffer and add a delay:

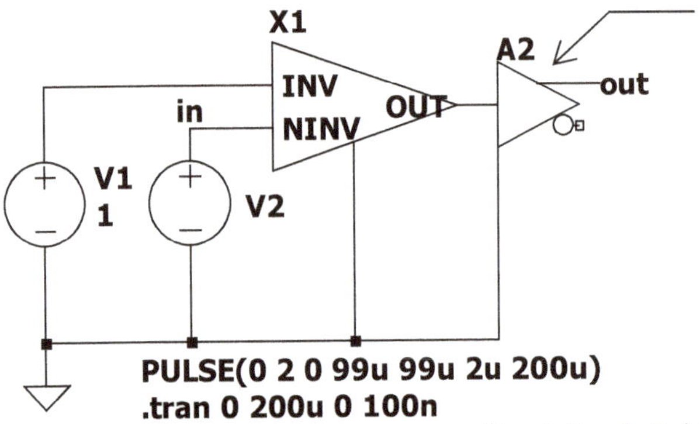

CTRL + right-click over the symbol

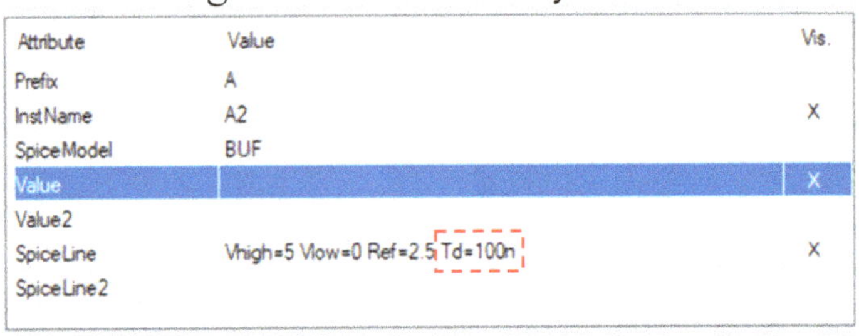

Attribute	Value	Vis.
Prefix	A	
InstName	A2	X
SpiceModel	BUF	
Value		X
Value2		
SpiceLine	Vhigh=5 Vlow=0 Ref=2.5 Td=100n	X
SpiceLine2		

test comparator with delay.asc

Adding Dead Time

DEAD TIME is an important parameter in different applications such as synchronous rectification (SR) or half-bridge (HB) circuits. Adding dead time between commutations, prevents overlap between the transistors and eliminates shoot-through currents. In some half-bridge applications, a dead time is mandatory to let the magnetizing current discharge the capacitors seen at the bridge node and bring zero-volt switching (ZVS).

The below circuit is a simple implementation using a *RC* network and can be built on the bench with a few logic gates. The time constant is calculated according to a 2.5-V threshold level for the adopted gates:

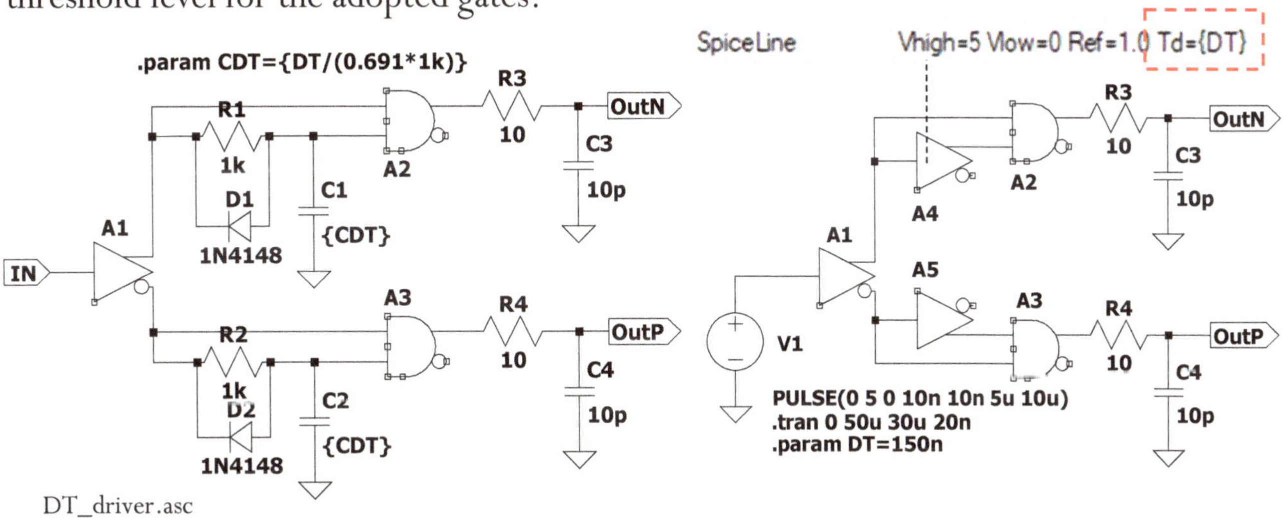

DT_driver.asc

A variant of this circuit exists in which a logic delay replaces the *RC* networks. The OutN node drives the upper-side MOSFET while OutP controls the low-side.

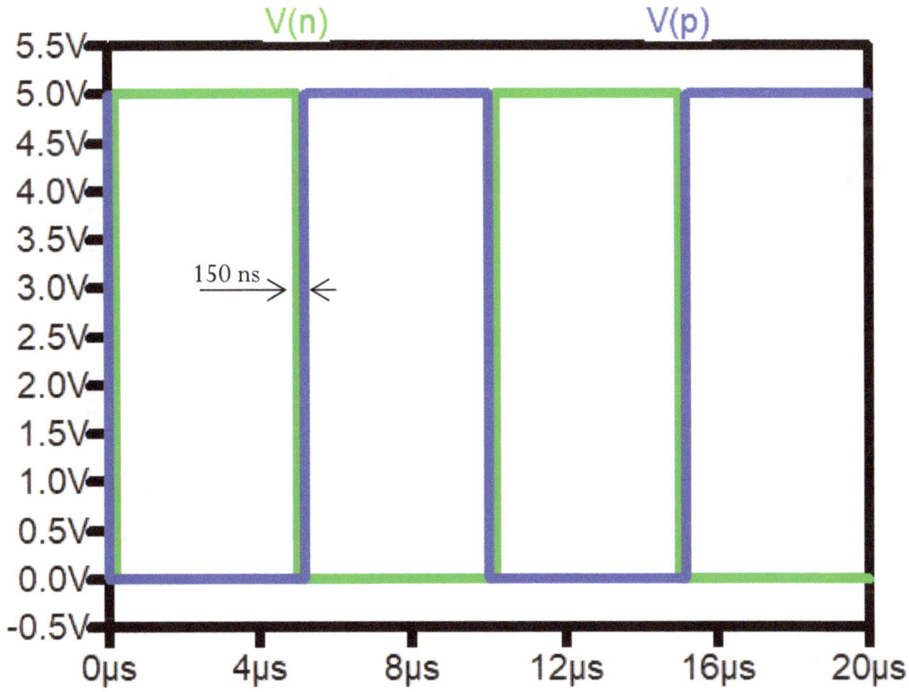

The dead time is 150 ns in this example. The maximum time step is reduced to 20 ns to offer adequate granularity for plotting the waveforms.

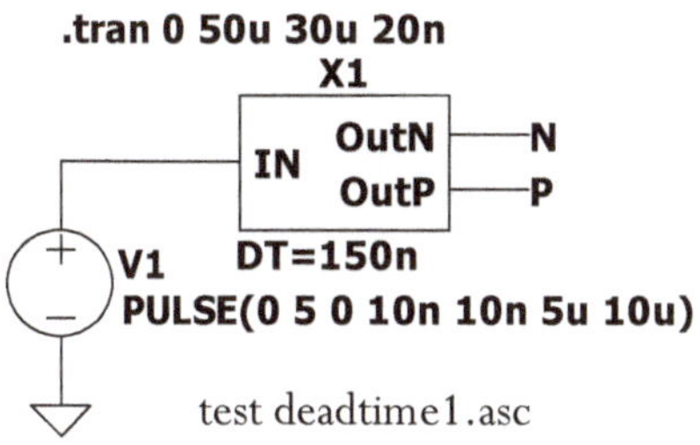

test deadtime1.asc

Small-Pulse Generator

GENERATING A small pulse synchronized with the rising- or falling-edge of a signal, is often necessary when building sampling circuits, leading-edge blanking (LEB) gate signals and so on. Two simple AND gates are used for this block. The first one carries the delay while the second one delivers 5 V only when its two inputs are high.

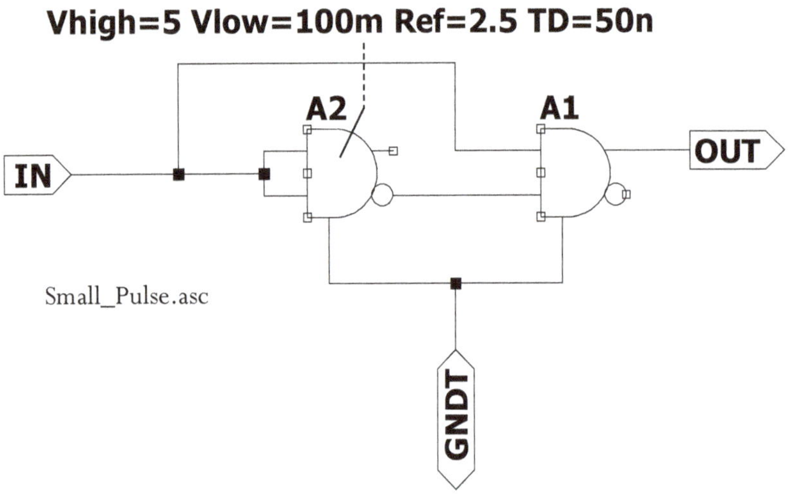

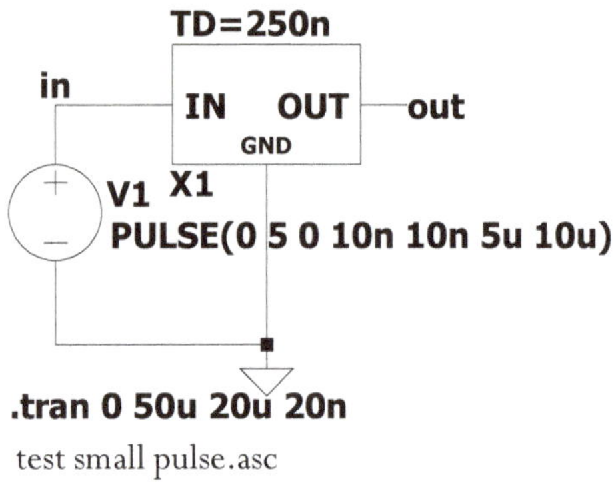

With this circuit, I can easily assemble a LEB circuit as classically found in modern current-mode controllers.

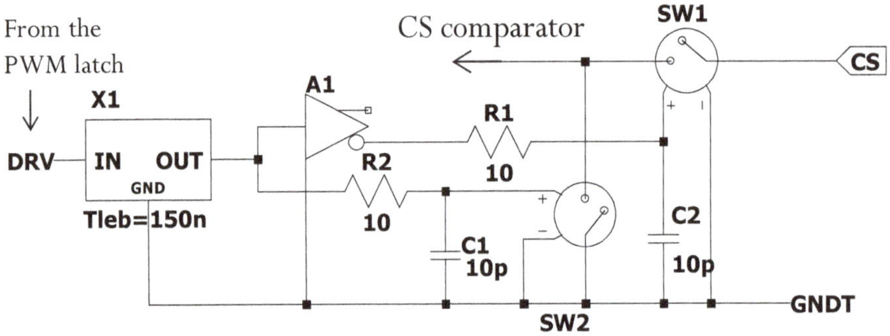

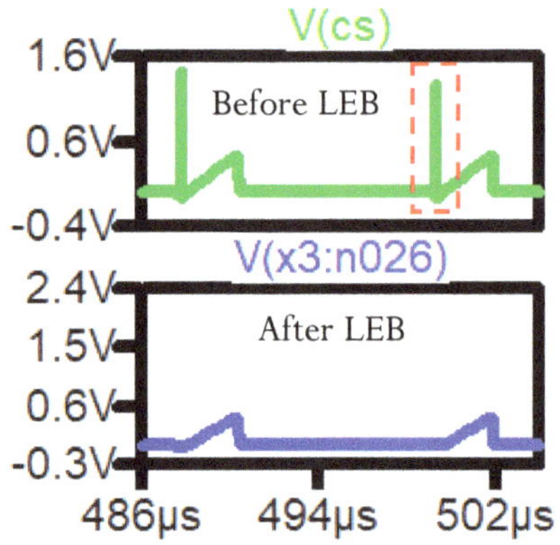

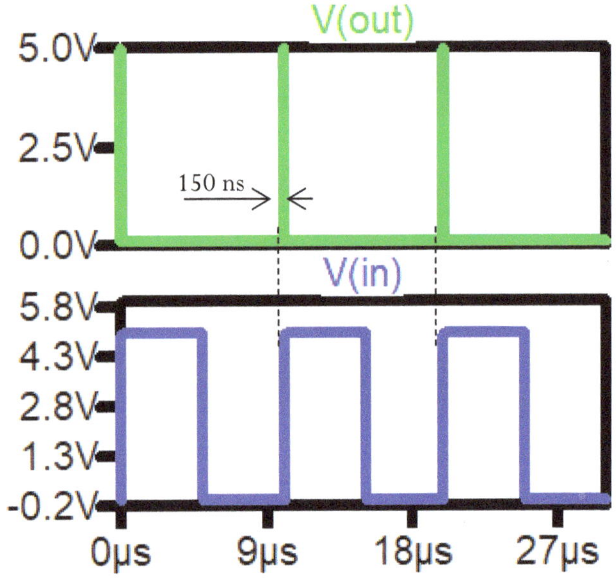

When the drive output (DRV) goes high, switch SW2 turns on while SW1 remains open: no signal appears at the current sense (CS) comparator. When the LEB time elapses, e.g. after 150 ns, the output of subcircuit X1 goes low. SW2 now opens and SW1 closes. A clean current-sense information is now conveyed to the comparator for observation and reset. Sometimes, controllers implement two distinct LEB circuits for building a two-level over-current protection (OCP).

A Digital Delay

THE DELAY we have inserted in the previous examples, belong to the time-domain analysis and can be of any value. In the sampling- or discrete-time world, a delay is an integer associated with a sampling circuit, the computational time of an ADC, and so on. Designated as z^{-1} in discrete-time equations, we need to find an equivalent circuit if we want to simulate digital compensators such as a type 2 or 3 with analog subcircuits. The easiest approach implements a delay line as described in [2]. It is easy to assemble and simulates fast in ac analysis. Unfortunately, the simulation suffers in transient analyses and I prefer resorting to an analog version of this delay, which has a good accuracy compared to the original delay line.

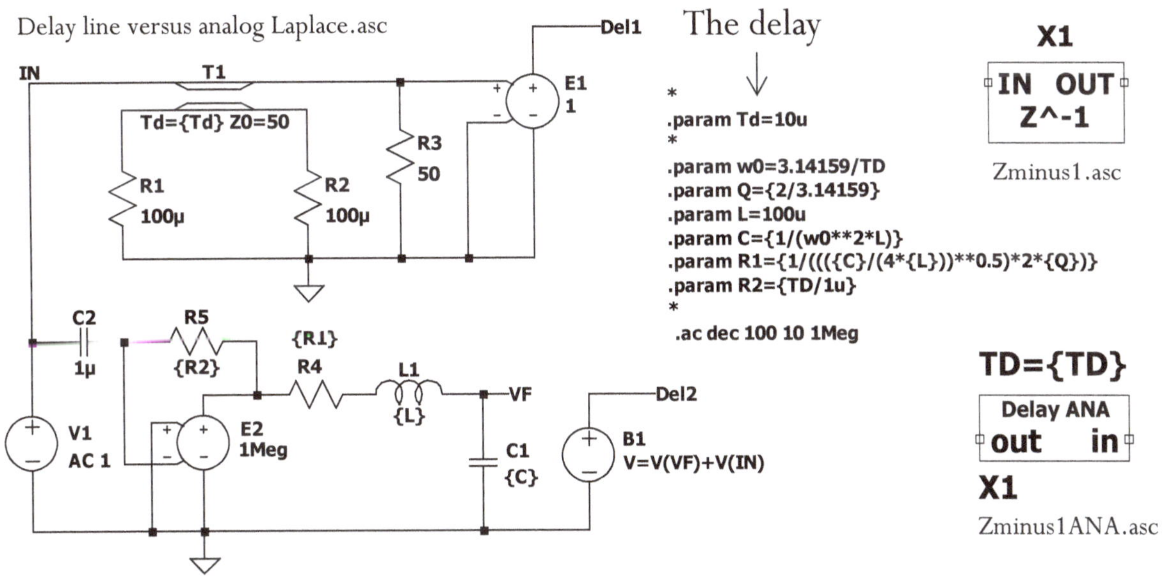

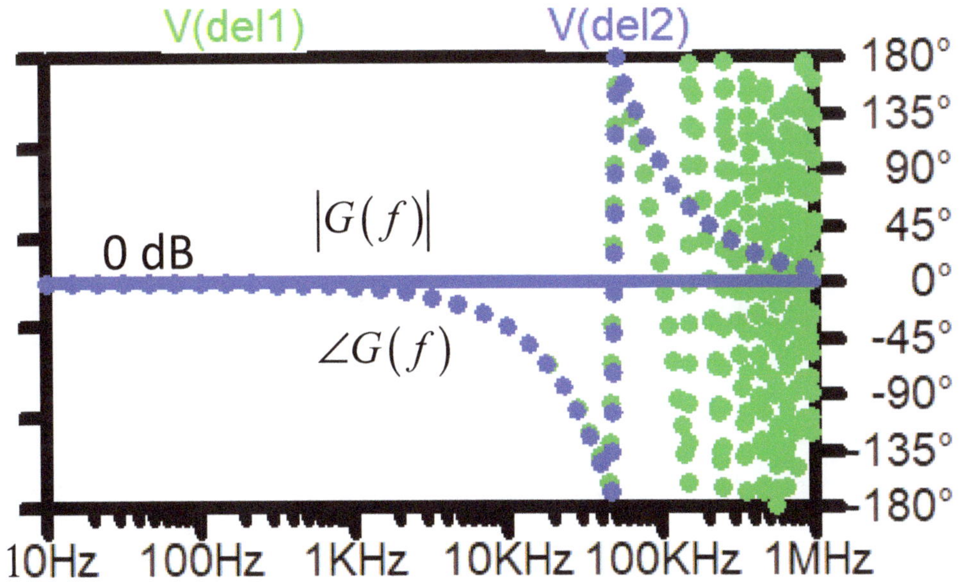

The ac responses of both circuits are very similar, up to the sampling frequency (100 kHz). The transfer function of the analog block is given below:

$$H(s) = 1 - \frac{s \cdot \tau_d}{1 + \frac{s}{Q\omega} + \left(\frac{s}{\omega}\right)^2}$$

$$Q = \frac{2}{\pi} \quad \omega = \frac{\pi}{\tau_d}$$

Generic Optocoupler Models

THE OPTOCOUPLER is a very common component for closing the loop in an isolated converter. The photons emitted by an LED inside a molded package, are collected by the base of a transistor, whose collector and emitter connections are externally accessible. By modulating the LED current, you can change the collector or emitter current, without having an electrical link between the two parts: you obtain *galvanic* isolation between the secondary (which drives the LED) and the primary, where the live switching parts reside.

You can find new techniques for isolating the loop that are even integrated with the control section. Magnetic and capacitive coupling are the possible options you will find on the market. The optocoupler, however, is still difficult to beat in terms of cost.

The part lends itself well to a control via a TL431 but you *must* characterize its parasitic pole prior to including it in the loop. This step is important, otherwise the internal parasitic capacitor may disturb your compensation strategy, affecting stability margins in worst cases.

The difficulty is to model the LED because, depending on where you operate the optocoupler – the bias current in the diode – its dynamic resistance r_d will change and can affect the gain, especially if r_d becomes significant compared to the series resistance R_{LED}. The CTR is also very dependent on these operating conditions and may vary a lot.

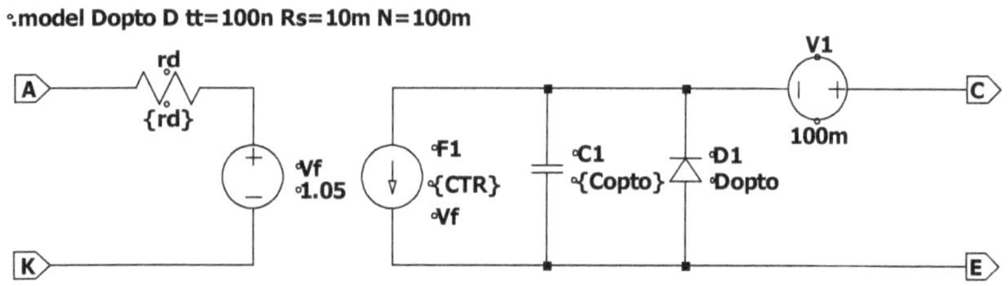

optoTOF.asc

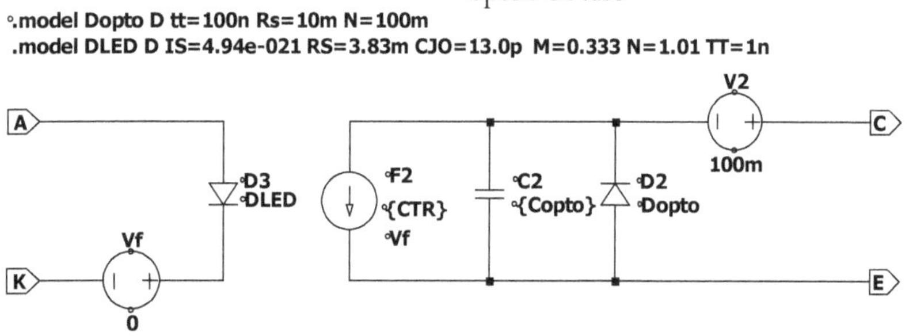

optoTOF_2.asc

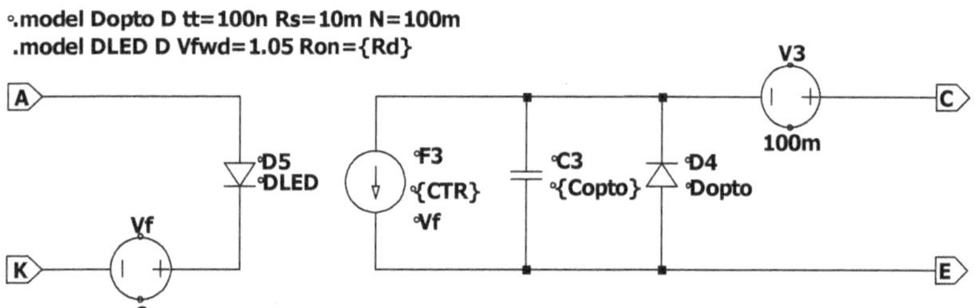

optoTOF_3.asc

In the simplest approach, the diode is modeled by its voltage drop V_f and a resistance mimicking r_d. Then you can replace this source by a more refined LED model in which the dynamic resistance will vary depending on the bias current. Finally, the third model resorts to a perfect diode in which you set the threshold and the resistance.

Make sure $\{C_{opto}\}$ is well characterized on the bench, this is important for stability.

Characterizing the Optocoupler

THE OPTOCOUPLER brings its own ac response, with multiple poles in the graph. However, considering crossover frequencies around a few kHz, it is acceptable to consider the optocoupler as a first-order system as we did in the previous page. When you have selected the right component – based on availability, isolation voltage, creepage distance, CTR, cost and so on – it is important to extract the pole and determine the equivalent collector-emitter parasitic capacitance, C_{opto} in the LTspice automated macros.

The below circuits are possible options to perform this test. The one with the coupling capacitor perhaps lends itself well for a measurement on the bench, while the second one is more appropriate with a frequency-response analyzer. In both cases, the LED dc bias must be adjusted to bring the transistor in its linear range, e.g. ≈2-3 V with a 5-V V_{dd}.

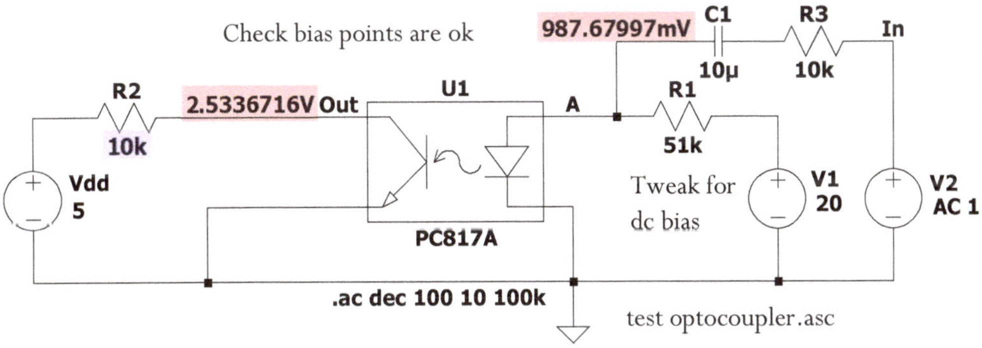

An oscilloscope can be used for extracting the pole on the bench: set a low-frequency sinusoidal waveform for V_2 (e.g. 100 Hz) and observe the sinusoidal response at the *Out* node. Carefully adjust the amplitude (keep the system linear) and the Y-scale on the oscilloscope to cover 10 vertical divisions. Now increase the frequency and check when node *Out* drops to ≈7 vertical divisions: this is your pole position at -3 dB (0.707).

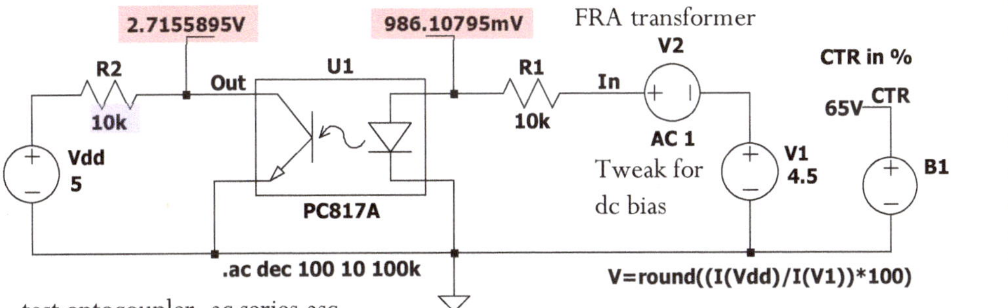

test optocoupler -ac series.asc

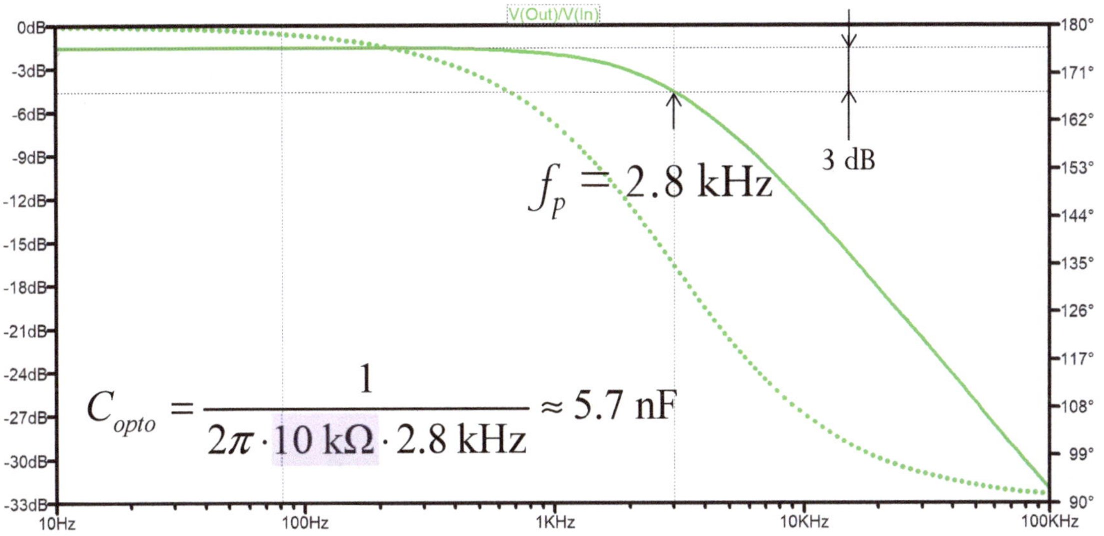

With the pole in hand, you can calculate the parasitic capacitance C_{opto} which depends on the pull-up resistance.

TL431 Model

THE TL431 is truly a marvelous 10-transistor, all-in-one reference voltage and operational amplifier housed in a 3-pin package. It is found in a myriad of ac-dc and dc-dc converters for regulating the output voltage. With its 2.5-V internal reference voltage, it is easy to hook an optocoupler to its cathode for nicely modulating the LED current. However, it is important that you understand the slow- and fast-lane concepts as compensating a TL431 differs from working with a classical op-amp. I will detail design examples of the three compensator types around the TL431. I have assembled a generic version of the part, for easy-converging simulations, but a discrete version is also available for both the TL and TLV431 (1.25 V reference, *TLV431TOF_DIS.asc*).

```
.model DTL431 D tt=100n Rs=10m N=100m Cjo=10p
.model DTL431CLP D tt=100n Rs=10m N=100m Cjo=10p BV=37
.model Q2N3904 npn ( IS=2.48E-13 VAF=101.7 BF=400 IKF=0.0334 NE=1.5243 ISE=2.113E-12
+IKR=0.02 ISC=5.00E-12 NC=1.1 NR=1 BR=10 RC=0.5 CJC=3.50E-12 FC=0.5 MJC=0.25 VJC=0.7 CJE=4.50E-12
+MJE=0.33 VJE=0.75 TF=2.80E-10 ITF=0.4 VTF=2 XTF=10 RE=1 TR=8.00E-07)
```

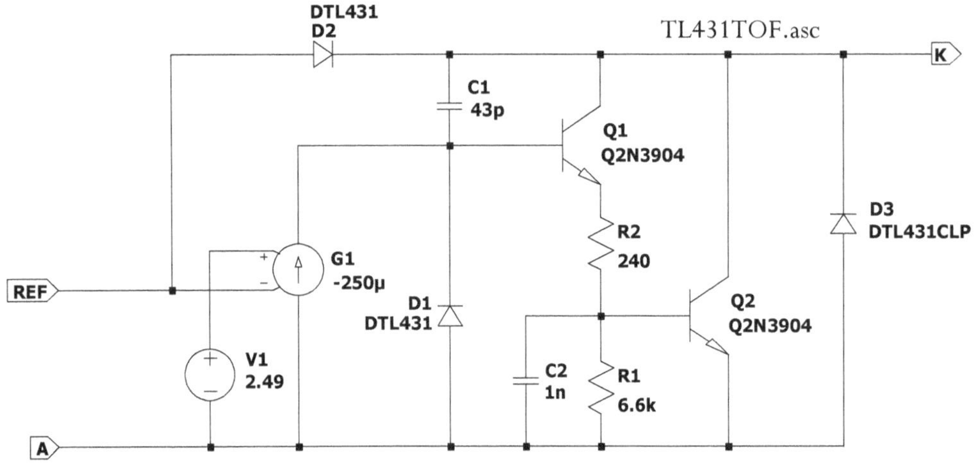

A voltage-controlled current source drives a series of transistors to offer the shunt function of the TL431. G_1 starts sourcing current to Q_1 when the REF pin exceeds the 2.49-V reference voltage. This activates Q_2, which pulls the cathode K down. C_2 and the associated resistances, shape the ac response of the part.

I have used the above model for decades in many different projects and it gave good results for stabilizing the loop on simulation, later confirmed by bench measurements.

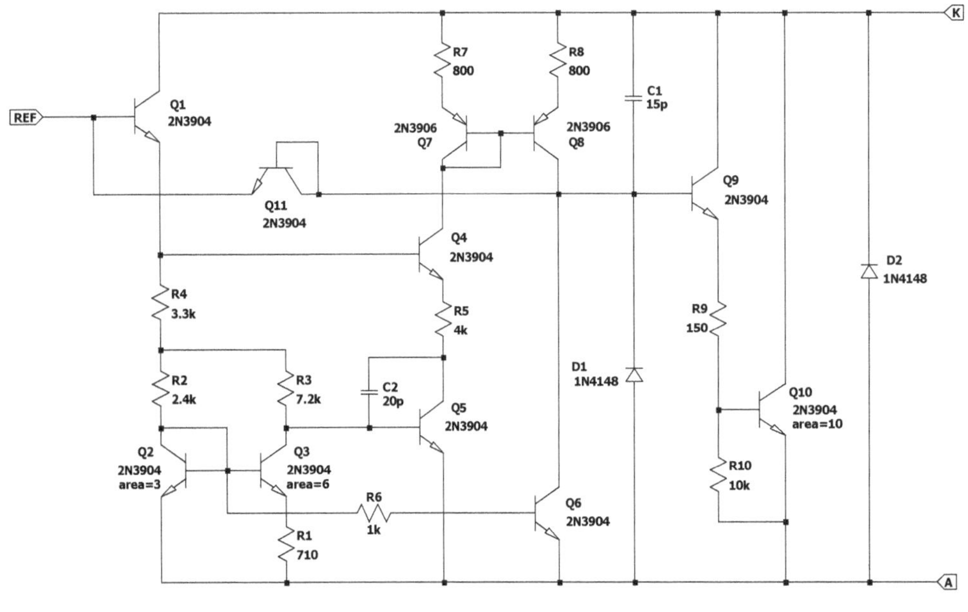

TL431TOF_DIS.asc

This is the discrete implementation of the TL431 using bipolar transistors. I remember an analog designer during my time with MOTOROLA, who told me that the 2N3904/06 were rather good discrete equivalents of the transistors found in a classical low-voltage bipolar process. He was using them for breadboarding some of his IC projects.

A Voltage-Controlled Oscillator

THIS SUBCIRCUIT can be implemented in an LLC converter when the control variable is the switching frequency. It could also be part of a frequency-foldback scheme, now commonly found in modern PWM controllers. There are plenty of ways to build a VCO and the below implementation is one of them. You can set the lower and upper switching frequency limits, but also the dead time applied to the half-bridge. The control is done via the pulldown of the feedback pin, with an optocoupler for example.

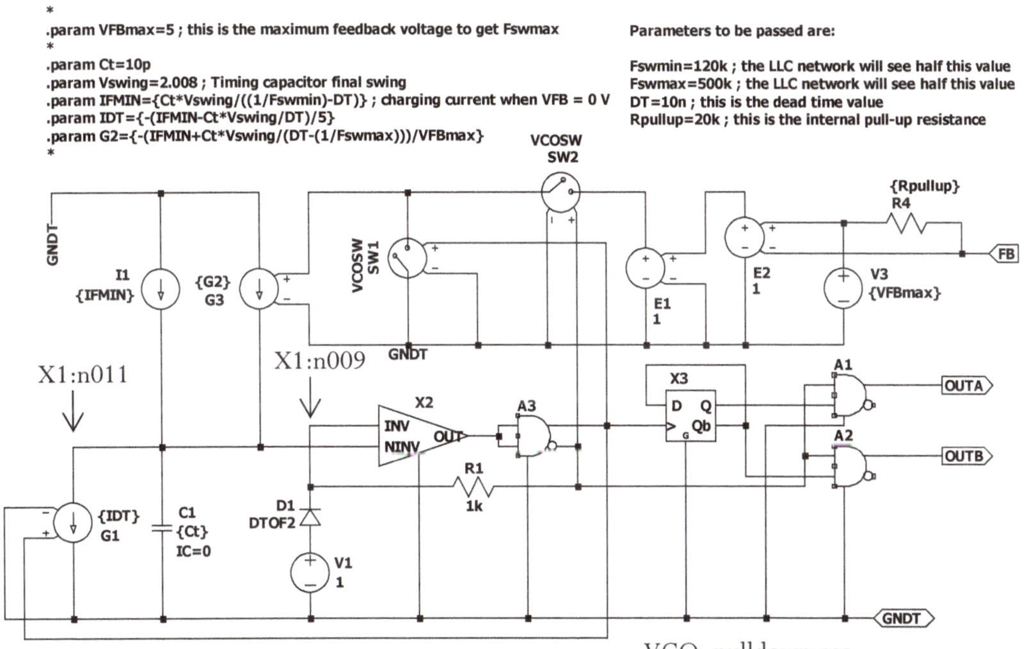

The minimum switching frequency is ensured by charging capacitor C_1 via the fixed current source I_1. When the voltage reaches 3 V (D_1 breaks down at 2 V), comparator X_2 toggles and discharges C_1 down to 1 V with G_1. Its transconductance sets the current for the wanted dead time.

VCO_pulldown.asc

In the below example, you can see how the C_t capacitor voltage swings between 1 and 3 V. Its discharge time is calibrated for a 150-ns dead time in this example. This VCO will later be used to control a high-power LLC converter.

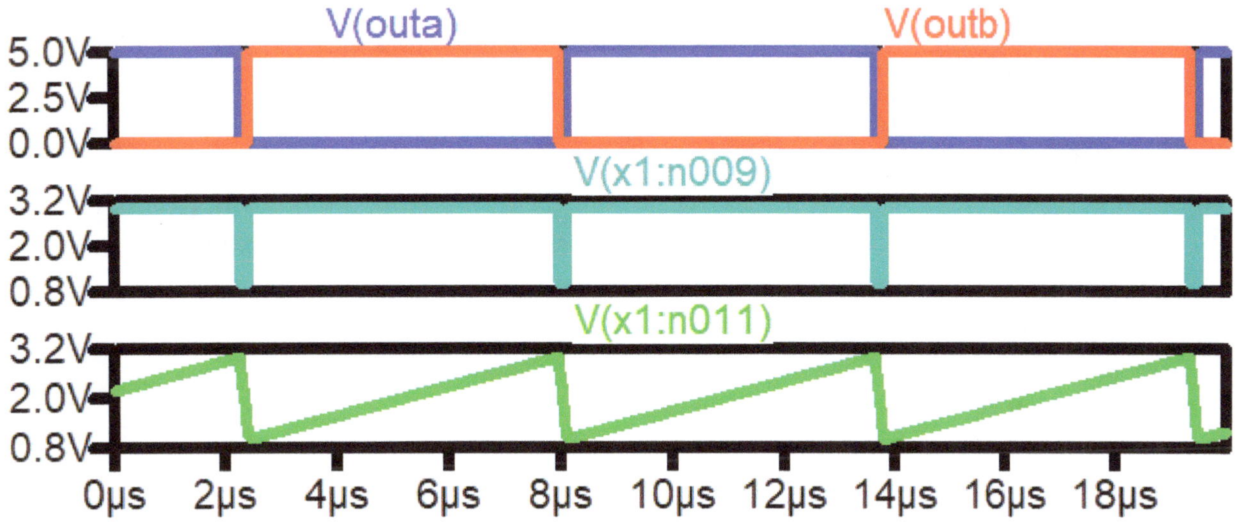

test VCO.asc

A Voltage-Mode Controller

NOW THAT WE have studied all these individual circuits, we can review a few of the generic controllers that I have assembled for these templates. We can start with the first one, a voltage-mode controller that is illustrated below.

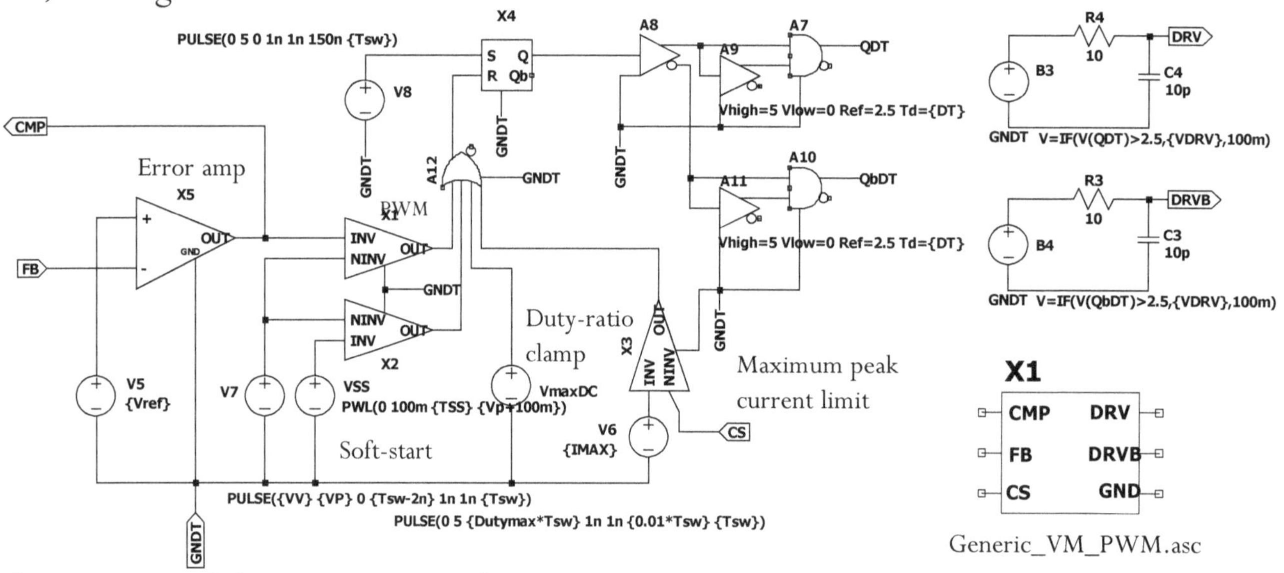

Generic_VM_PWM.asc

The error amplifier section uses the generic op-amp I have already described. Its output can be pulled down to ground via an optocoupler if necessary. Its output feeds the inverting pin of the PWM comparator X_1 while the non-inverting pin receives a programmable sawtooth. Another comparator, X_2, also receives the sawtooth but its other input is biased by a voltage ramping from 100 mV to slightly above V_p, the sawtooth peak voltage. The duration of this ramp is T_{ss}, the soft-start time.

The main latch X_4 is set by narrow pulses programmed in V_8. Reset occurs either via the op-amp and X_1, by the soft-start comparator X_2 during T_{ss}, or also by the maximum duty-ratio limit set by V_{maxDC}. Finally, an overcurrent comparator also resets the latch when the voltage at the CS pin exceeds the programmed threshold, I_{max}. In this model, for the sake of simplicity and execution speed, latch X_4 turns off Q, but a safer option would be to fully disable the output gates A_7-A_{10} via another latch, also triggered by X_3.

In the upper right-side of the circuit, you will recognize the dead time generator, driving the two complementary outputs for synchronous rectification. This is very handy if you plan to build synchronous buck or boost converters. I have also used this model to test an active-clamp forward converter. If you choose the single output Q, just leave the other one unconnected (or pulled down to ground via a 1-kΩ resistance if convergence fails).

I did not include a leading-edge blanking circuit when reading the voltage at the CS pin, but it would be easy to integrate the LEB section I have described in the previous pages.

This VM controller will be used in many of the converters I will describe in the following pages, so it is important to understand how it operates.

A Current-Mode Controller

MANY OF THE SWITCHING converters you can find in consumer or industrial projects, operate in current-mode control. The control-to-output transfer function is well behaved – it is a 1st-order response at low frequency – and mode transition is seamless. However, slope compensation is necessary to calm down subharmonic poles located at $F_{sw}/2$, but it is a well-understood process these days. The circuit is shown below:

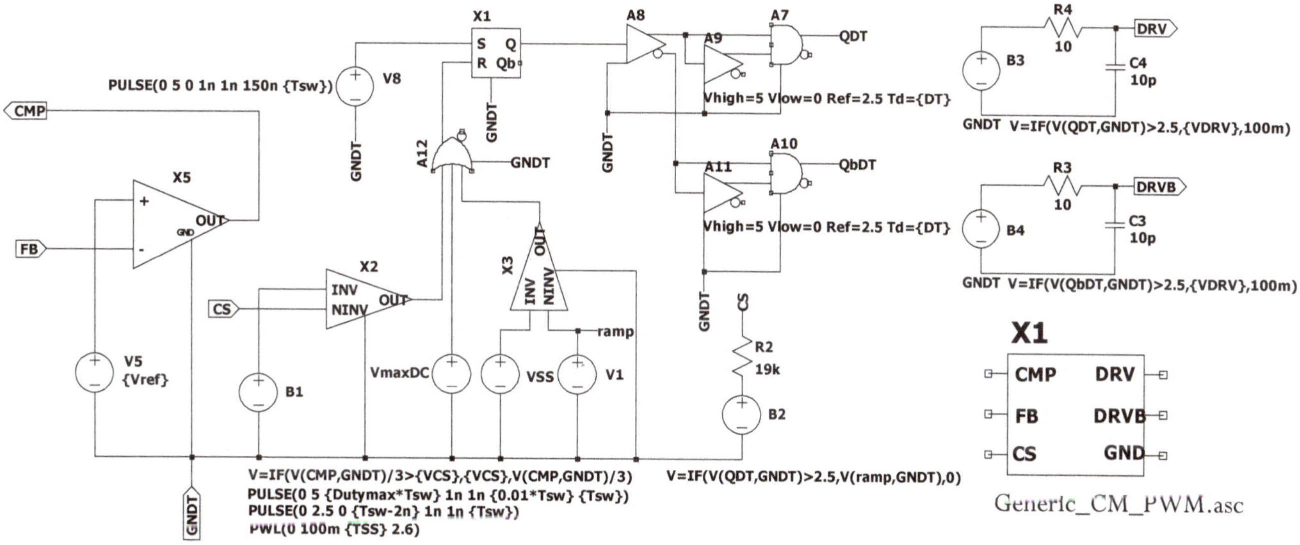

Generic_CM_PWM.asc

The error amplifier section does not change compared to the VM model. A clock still latches X_1 but the reset now comes from the current-sense comparator X_2. There is no delay in this block, but you can easily add an external gate to create one and match a datasheet parameter if need be. The V_{maxDC} source sets the maximum duty ratio in case of fault (shorted or open CS pin) or operational limit requirements like in a forward converter.

Ramp generator V_1, together with the V_{SS} source, offers a means to implement soft-start. The same ramp is used to inject slope compensation via a 19-kΩ resistance connected to the CS pin. By adjusting the value of resistance R_r carrying the voltage information from the sense resistance, you can easily tune the amount of slope compensation. That process will be automated by the macro in most of the templates.

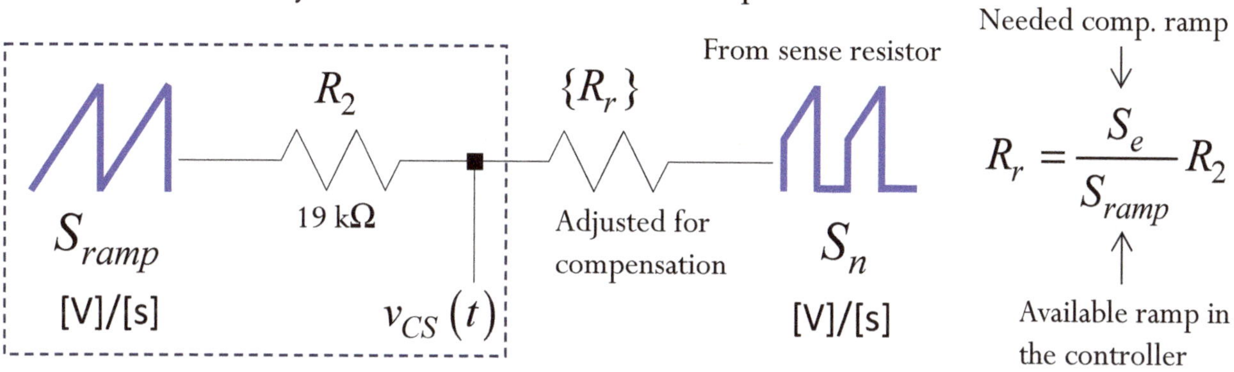

A Quasi-Resonant Controller

A QUASI-SQUARE-WAVE resonant converter – often abbreviated as QR – is a fairly common structure found in ac-dc adapters for notebooks or in consumer products. Rather than operating at a fixed switching frequency, the controller starts a turn-on sequence which ends when the peak inductor current has reached a threshold, set by the error voltage. The primary inductance depletes during the off-time and when the current hits zero, a new cycle is started. Often, a small delay is inserted before restart so that the switch drive is synchronized with the drain-source voltage reaching a minimum. This is called valley-switching operation. The frequency varies and depends on the load/line configuration. When the frequency is too high in light-load conditions, a VCO takes over and folds the frequency back to improve efficiency. By keeping turn-on synchronization with a valley in foldback, the VCO imposes discrete frequency steps and induces an instability known as *frequency hoping*. Depending on the transformer construction, this inherent instability can cause acoustic noise. A valley-lockout circuitry freezes the valley position for a given power level, and authorizes a change if the operating point evolves.

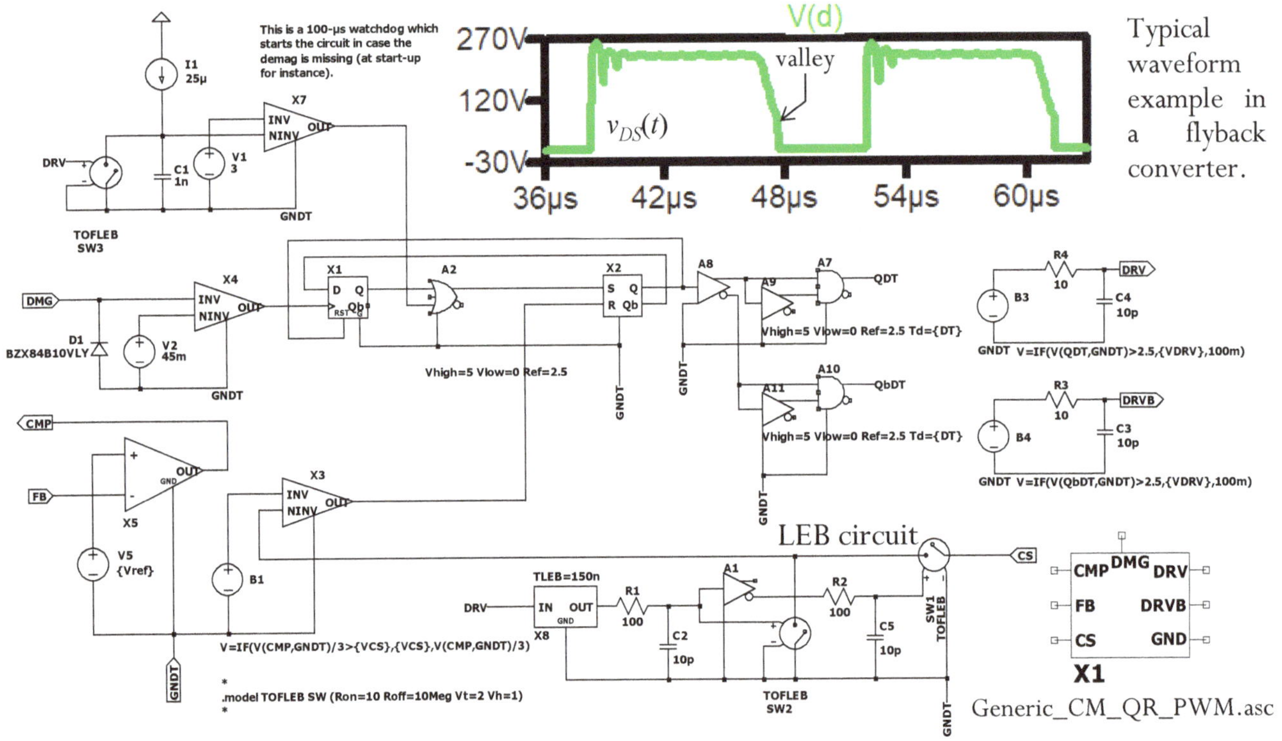

Typical waveform example in a flyback converter.

A new cycle is started when the voltage at the demagnetization pin (DMG) falls below 45 mV. In case this signal never comes – e.g. at start-up where the reflected voltage is low – a so-called watchdog wakes the circuit up and ensures low-frequency operation until V_{out} reaches a sufficient level after a few cycles. I added a LEB circuit this time, in the bottom of the circuit. As underlined in the introduction, there is no frequency clamp or foldback and the circuit will always operate in the first valley.

An Asymmetrical Half-Bridge Controller

THE ASYMMETRICAL HB controlled or AHB, is now gaining popularity owing to its high-frequency zero-voltage switching (ZVS) operation in high-density USB-PD adapters. The circuit looks very much like an LLC converter, but the secondary is that of a flyback converter. Operated in QR mode with the controller below, the AHB offers a safe operation of the power transistors and recycles the leakage inductance energy. In terms of components, it is the same quantity as in the active-clamped flyback, but it performs better and safer in my opinion. The key here – as in the ACF – is to force the magnetizing current to hit a negative value and have this current circulate in the bridge to ensure ZVS. In my model, a simple comparator monitors the magnetizing current – it is not possible in reality and timers are actually implemented in a dedicated controller – and releases the low-side switch when threshold I_{neg} is hit. This parameter is adjustable in my model and must be selected based on the total capacitance lumped at the HB node. The right amount of negative valley will bring ZVS in the upper-side transistor, while minimizing the rms content and core losses as operation is pushed in the third quadrant.

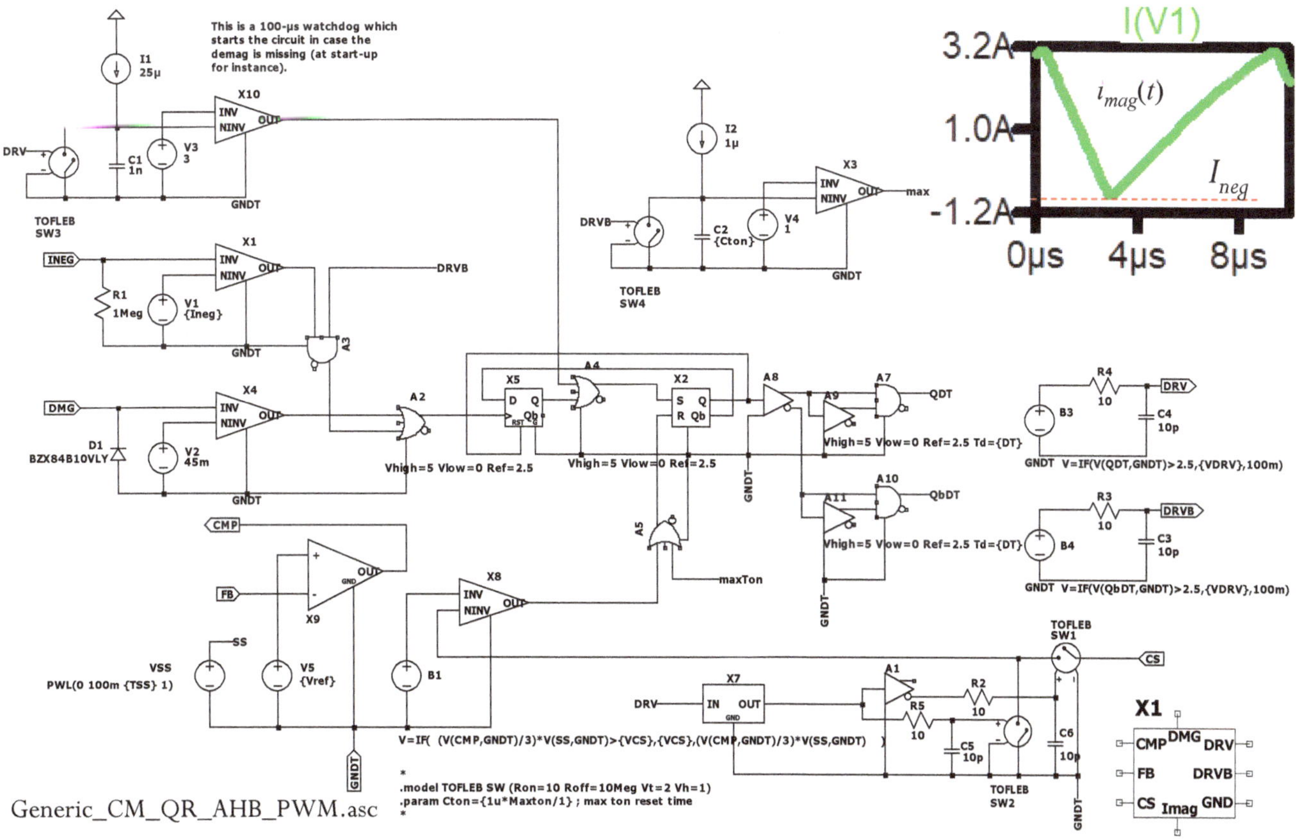

Generic_CM_QR_AHB_PWM.asc

Exactly as with the QR controller, a watchdog ensures the restart in case the demagnetization signal is too low or missing. Finally, a maximum on-time protection avoids driving the upper-side transistor for too long in case of a fault condition, like the absence of current-sense information. This duration is adjustable via the parameters.

Push-Pull Controllers

THE PUSH-PULL converter is often found in high-power dc-dc converters like in telecom bricks, for instance. In this type of configuration, the transformer operates in two quadrants (I and III), resulting in a better core utilization. I have prepared voltage-mode and a current-mode versions, depending on the application. Always check for possible flux walkaway in voltage-mode control and this paper [3] from TI brings interesting points. The model includes a soft-start that can be adjusted via the {Tss} parameter.

This is the current-mode controller which combines a voltage setpoint divided by 3 and an adjustable maximum sense voltage {VCS}.The circuit also includes slope compensation if the magnetizing current is not sufficient.

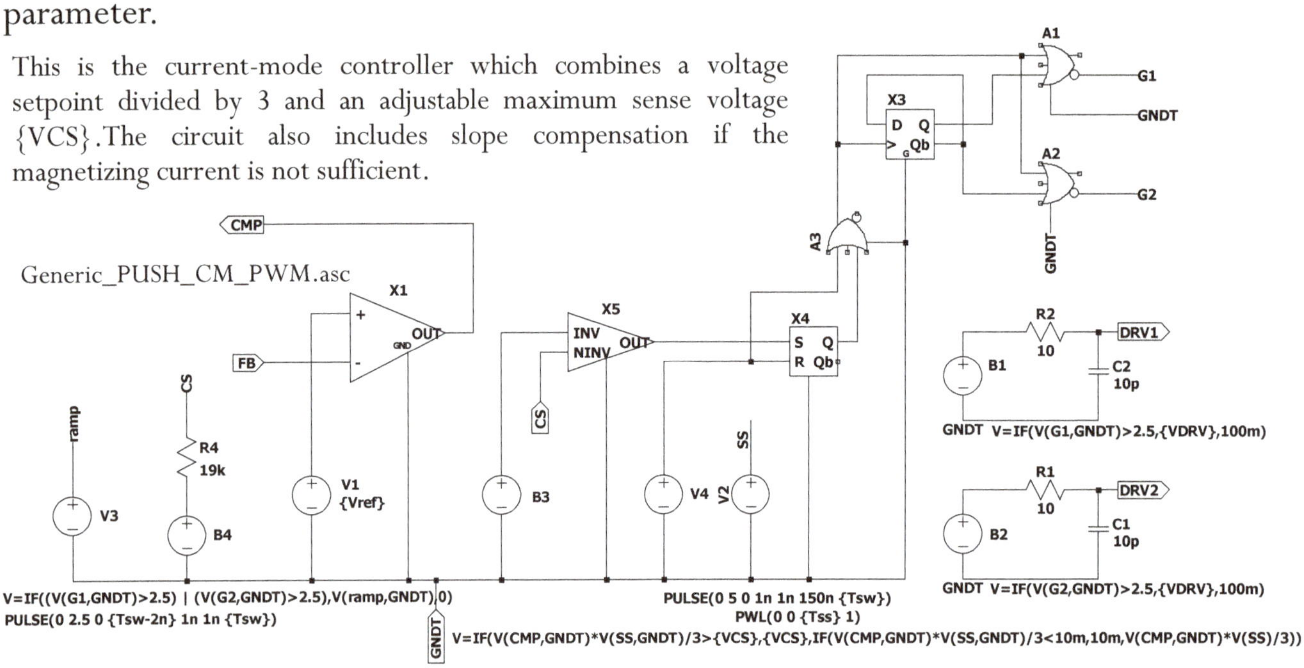

Generic_PUSH_CM_PWM.asc

The voltage-mode version is very similar in terms of implementation. The op-amp includes the 2.5-V reference voltage, whose output can be pulled down to ground for regulation in an isolated version. There is a cycle-by-cycle current limit which resets the PWM latch in case of an over-current, bringing both outputs down.

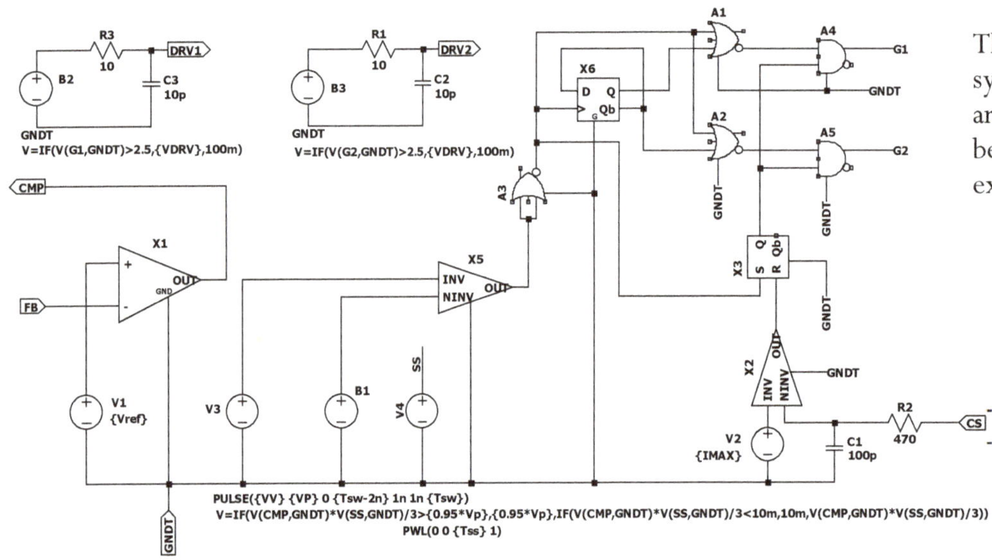

The transformer is symmetrically driven around zero as shown below. It ensures an excellent core utilization.

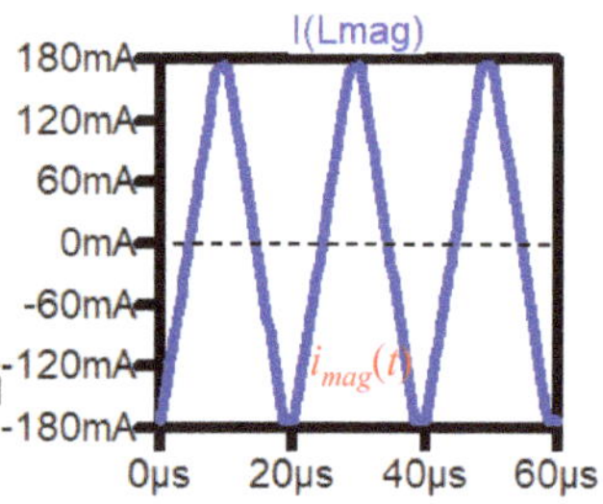

Generic_PUSH_VM_PWM.asc

Phase-Shifted Full-Bridge Controllers

THIS CONTROLLER requires more gates and flip-flops than with the previous controllers, considering the four switches to control. Both half-bridges are operated at a 50% duty ratio (with dead time). If fully synchronized, there is no voltage across the transformer and no power transfer. By shifting the phase between the two half-bridges, you have a means to adjust the power flow. Using the leakage inductance – which can be supplemented with an external inductor – the PSFB lends itself well for high-power applications as it brings ZVS in certain conditions.

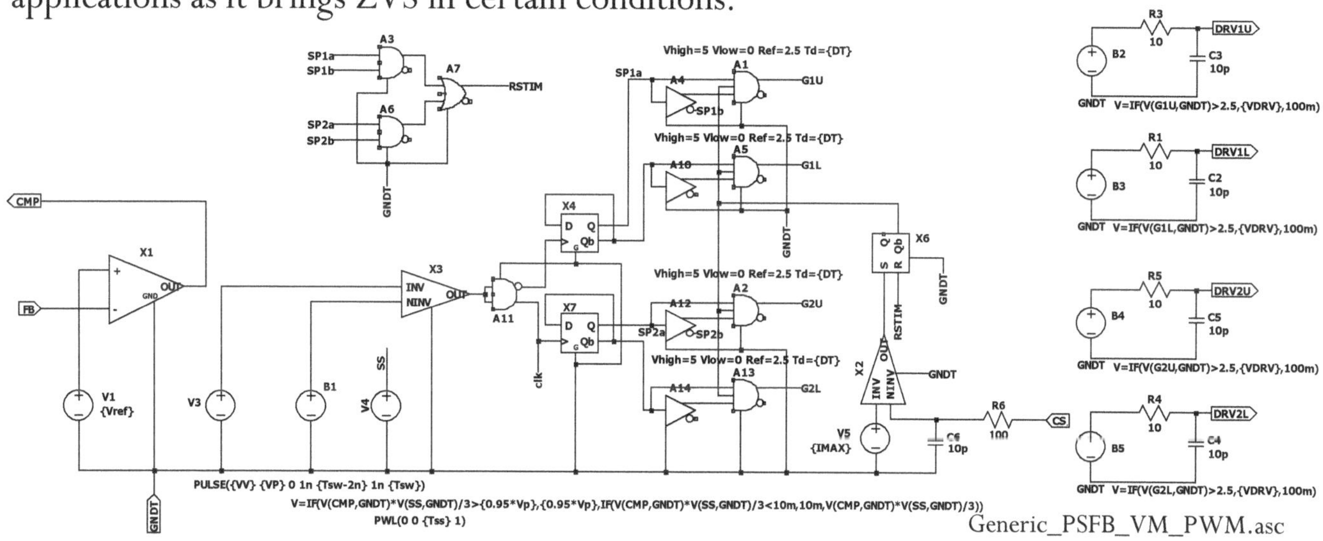

Generic_PSFB_VM_PWM.asc

In this voltage-mode version, the op-amp classically provides a voltage setpoint which is divided by 3 before driving the PWM block. A soft-start sequence via source V_4 and parameter $\{T_{ss}\}$ help reduce the stress at power up. The phase shift is created by X_4 and X_7, both clock inputs being driven by the PWM comparator. Finally, an over-current comparator shuts all 4 outputs down in case of over-current events. The CM version is very similar and requires an extra latch with X_3, as shown above.

The current-sense section could be supplemented with an upper fixed threshold (above V_{sense}) to shut down all switches as in the VM version.

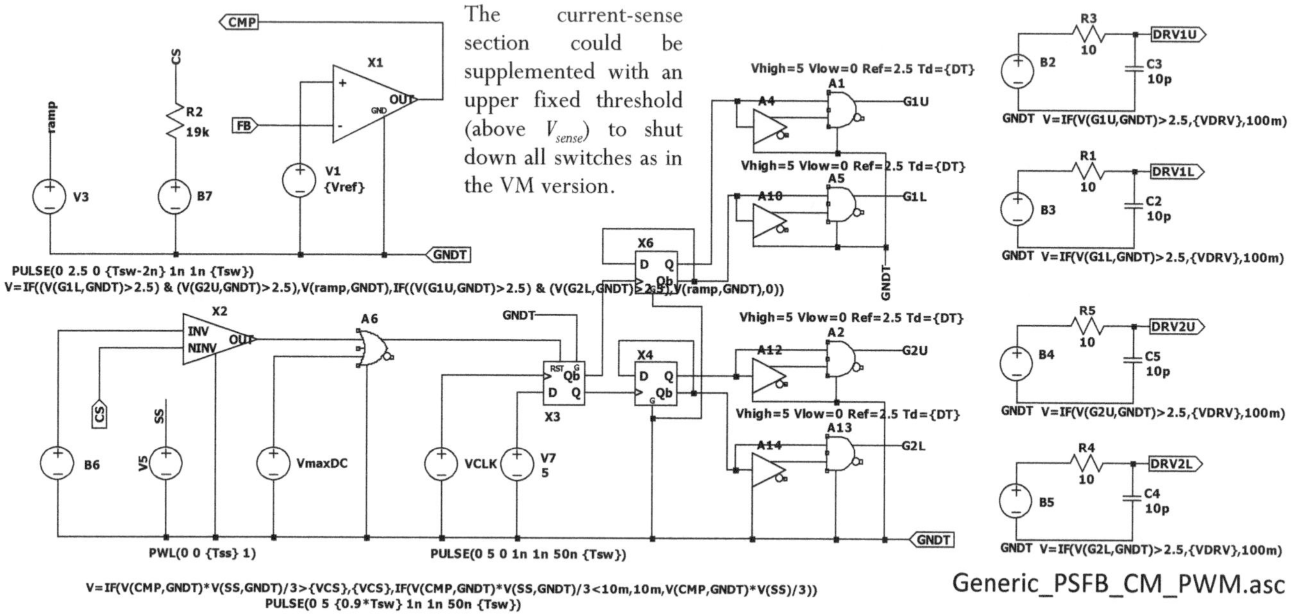

Generic_PSFB_CM_PWM.asc

A Dual-Active-Bridge Controller

THE DAB CONVERTER has gained popularity in high-power projects, like electric vehicles (EV) charging, in which bidirectionality is sometimes needed. The circuit uses a power transformer for scaling and isolation purposes, while an extra power inductor controls the power flow in both directions. Many types of control strategies exist and they help increasing the ZVS operation zones for the switches but also minimize the circulating rms content for reducing conduction losses. The most common technique is designated as single-phase shift (SPS) in which all half-bridges are operated at a 50% duty ratio. Dual- (DPS) and triple- (TPS) phase shift also exist and can be combined together, optimizing the efficiency along a wide operating range. The simple controller I propose brings SPS only as implementing DPS and TPS would seriously complicate the circuit. These options are illustrated at the bottom of this page.

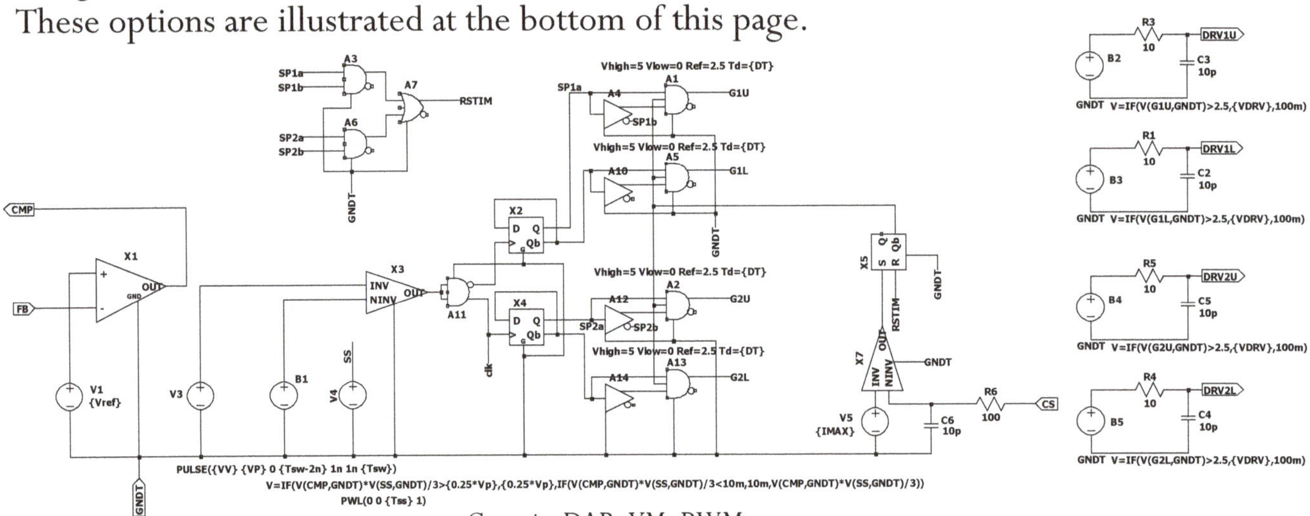

Generic_DAB_VM_PWM.asc

The logic blocks arrangement is very similar to that of the PSFB, except that a clamp is applied to the control variable (the angle φ) setpoint, to limit the angle below 90°, where the transmitted power peaks. This controller offers a unidirectional power transfer for the sake of simplicity. Considering the two full-bridges (one in the primary side and the other in the secondary), one would need eight different outputs. However, one output simultaneously drives two transistors, located in the primary and secondary. Isolation would of course be needed in a real converter, but for testing the principle of operation, it is not necessary. A dead time generator {DT} is added to provide adequate conditions for ZVS, while minimizing the conduction time in the 3rd quadrant for the transistors.

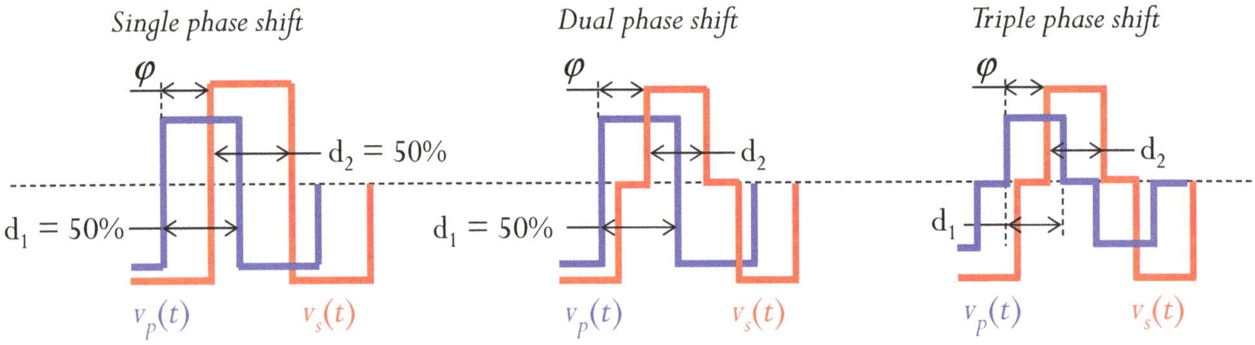

An LLC Controller - VM

A SIMPLE LLC controller can be derived from the VCO circuit presented in the previous pages. I have added a couple of *B* sources to offer the ability of controlling a high-side source with the SU node connected to the high-voltage half-bridge (HB) node. The parameters expected for the model are the following;

- F_{swmin} and F_{swmax} set the minimum and maximum switching frequencies respectively. Please note that these values are divided by 2 at the HB node.
- DT sets the dead time between the upper- and lower-side transitions
- R_{pullup} sets the internal resistance internally hooked to a 5-V V_{dd} in the subcircuit.

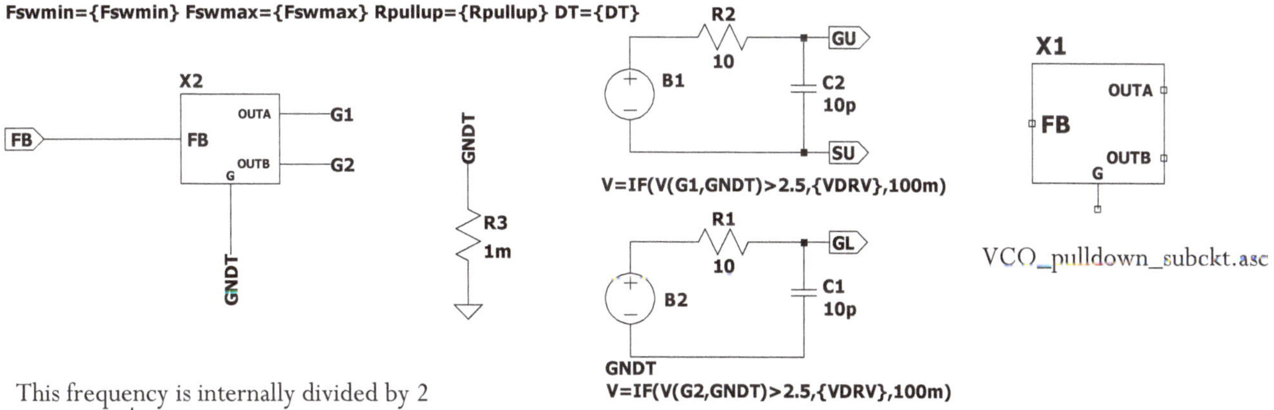

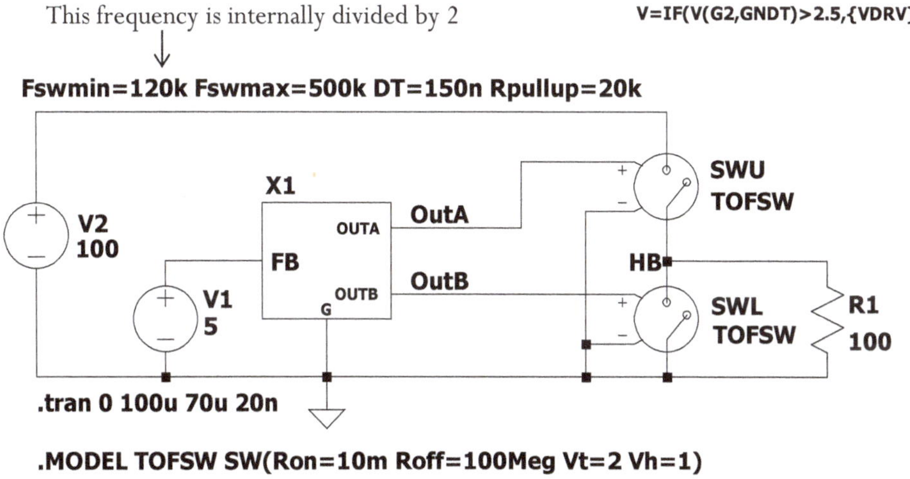

In this example, the feedback node is held high, pushing the switching frequency to its minimum value, which occurs during start-up or when the loop is broken: the VCO pulses at F_{swmin} in this mode. When the FB pin is low, the switching frequency is maximum and would be 250 kHz in this example.

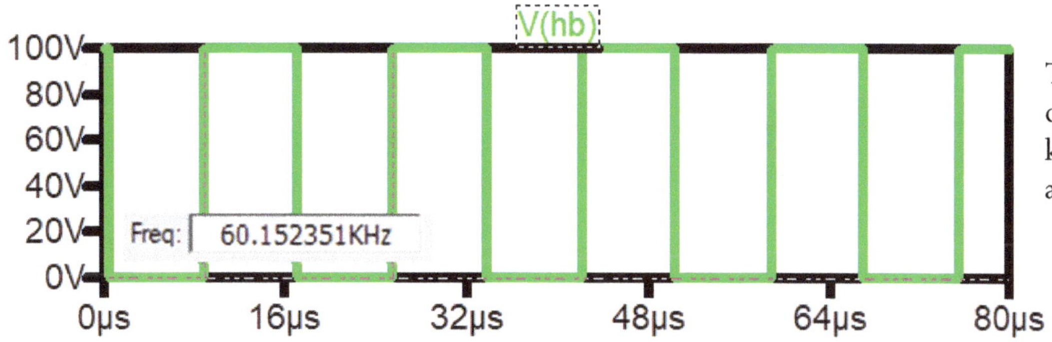

The simulation confirms a 60-kHz frequency, as expected.

An LLC Controller - CM

THE DIRECT-FREQUENCY-CONTROLLED (DFC) LLC operates with a VCO setting the switching frequency *directly* from the error voltage. It is the same as with a voltage-mode-controlled converter where the control variable – the duty ratio D – directly relates to the error voltage via the pulse-width modulator. In current-mode control, the error voltage sets the inductor peak current which, *indirectly*, sets the duty ratio D. If D is identical in a converter operated in VM or CM, the dynamics drastically change between the two modes. The same analogy exists with the LLC, except that the control variable is the switching frequency F_{sw}. If the error voltage now drives the primary peak current, we have a means to *indirectly* control the switching frequency and enjoy a much better frequency response.

CM_modulator_subckt.asc

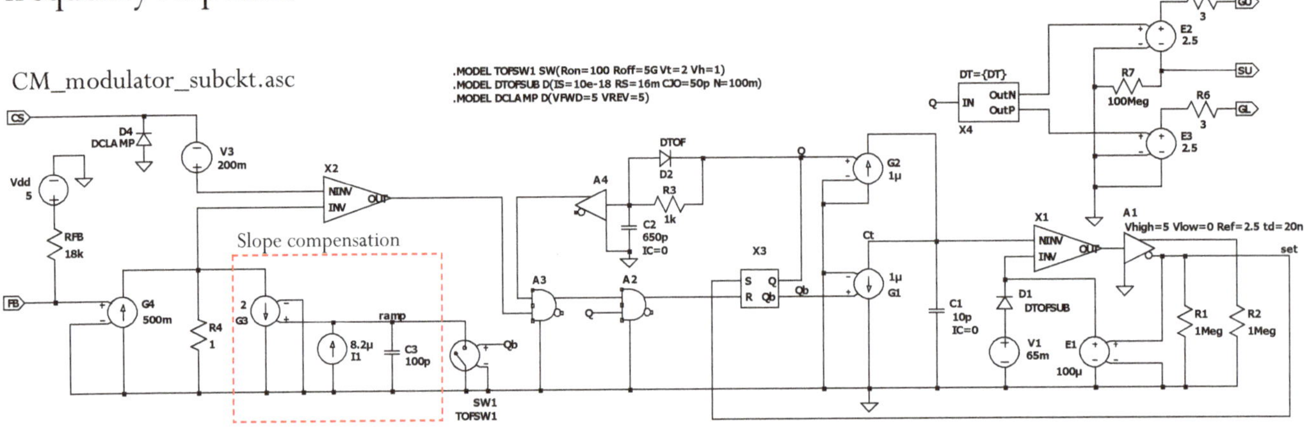

Above is a simplified modulation section of a current-mode controller, like the NCP13992 from *onsemi*. The principle is to generate the on-time of the upper switch via capacitor C_1 and duplicate this time period via the discharge of the capacitor down to low level. By doing it this way – it is digitally implemented in the LLC controller – we create a 50% duty ratio switching pattern whose frequency will adjust based on the operating conditions. The control-to-output transfer function approaches that of a 1st-order system.

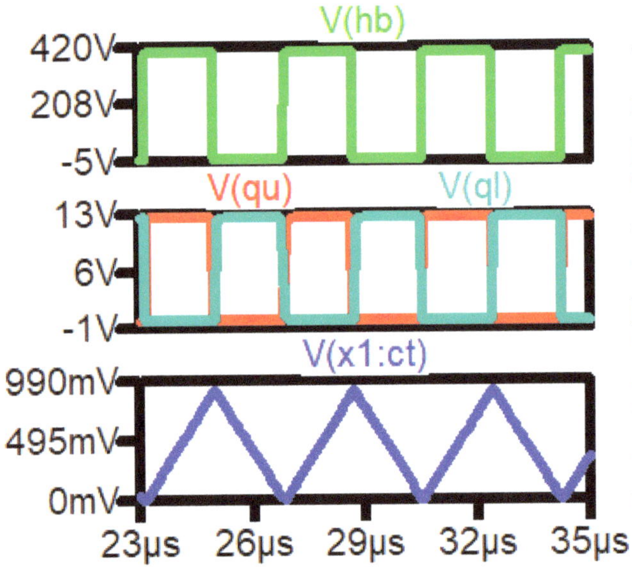

This current-mode model requires slope compensation for proper stability and all is internal to the controller. In the simulation graph, you can see the driving waveforms with a programmed dead time. Capacitor C_1's upper limit depends on the wanted on-time. The off-time will replicate this value for an exact 50% duty ratio.

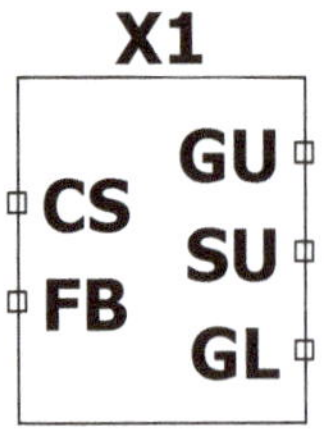

An LLC Controller – Bang-Bang Control

RATHER THAN CONTROLLING the peak primary current, it is possible to observe the peak and valley voltages of the resonant capacitor instead. From the valley to the peak, energy is stored and then transmitted to the secondary from the peak to the valley. By symmetrically controlling these voltages, you can adjust the power flow and, *indirectly*, set the switching frequency: this technique is called bang-bang charge control. Again, an LLC converter driven in DFC, CM or bang-bang charge control, will operate at exactly the same frequency in identical line/loads conditions. However, its transfer function will be well behaved in the two last cases, authorizing a simpler and more efficient control strategy.

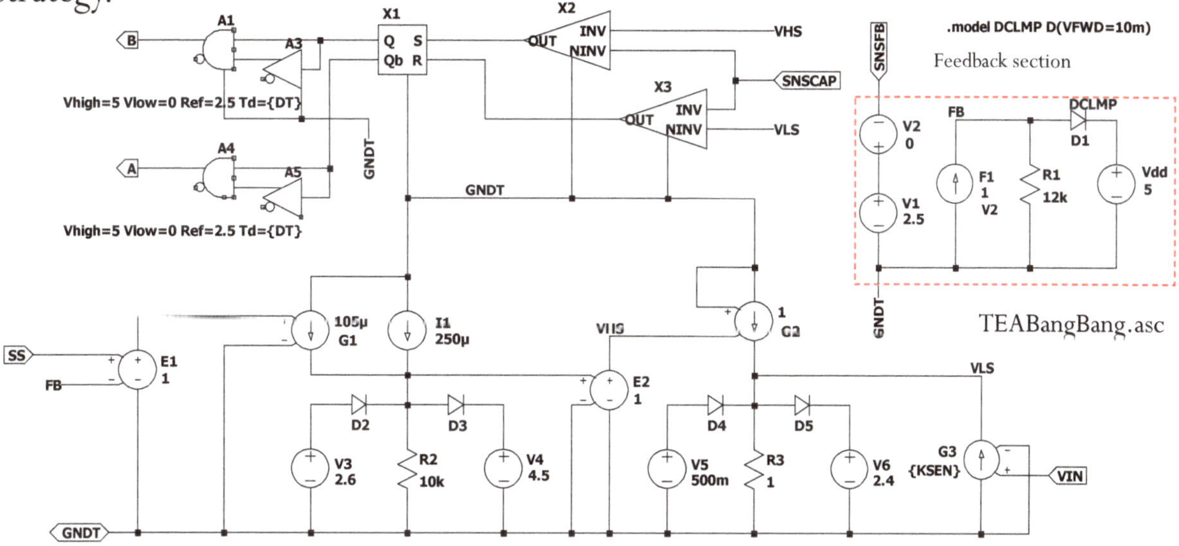

TEABangBang.asc

Above is a simplified modulation section of a bang-bang controller such as the TEA2226AT from *NXP*. The optocoupler sinks current from the FB pin and changes the upper and lower levels, respectively labeled VHS and VLS. These nodes are routed to comparators X_2 and X_3, which receive an image of the voltage across the resonant capacitor. In the real implementation, that voltage is brought to the IC pin via a compensated resistive divider: this is a high-impedance path and you will have to carefully route it to the IC.

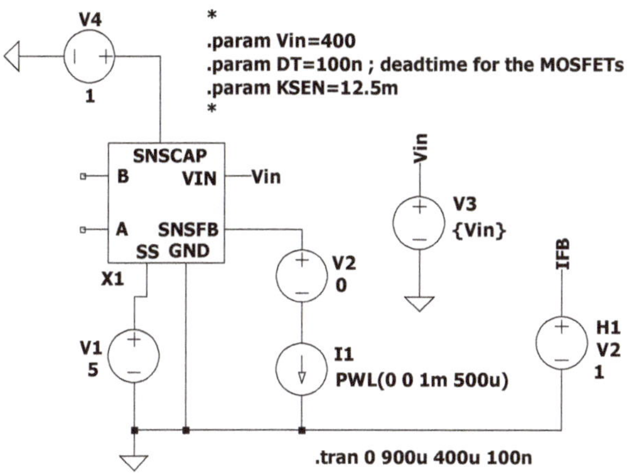

test bangbang modulator.asc

You can clearly see how drawing current from the FB pin, drives down the peak and valley voltage setpoints, adjusting the power flow.

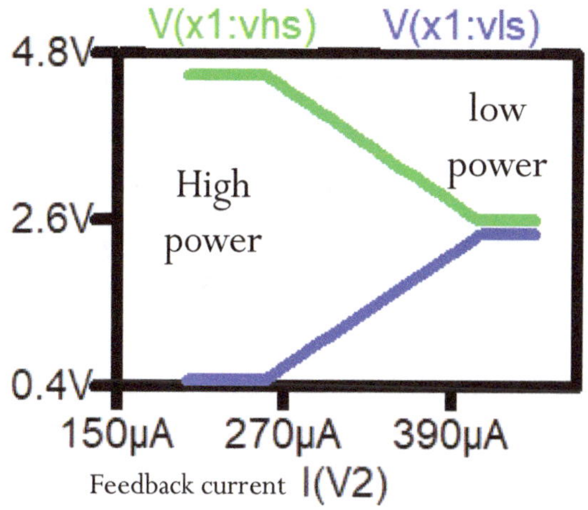

The UC384X Controllers

THE UC384X is a generic part-number, describing a family of popular current-mode PWM controllers introduced by Unitrode (now Texas Instruments) in the 80s. Processed with the bipolar technology, the 8-pin integrated circuit is available in different flavors of under-voltage lockout (UVLO) and duty ratio limits. For example, the UC3842/43 offer a maximum duty ratio of 92% (min.) and are available with different on and off V_{cc} levels. They are well suited for topologies accepting these large duty ratio swings like a flyback operating on universal mains or a buck converter, provided you add the adequate driver externally for the latter. On the other hand, a forward converter featuring the classical 1:1 auxiliary reset winding, cannot accept more than 50% of maximum duty ratio, and the UC3844/45 are appropriate for this purpose. The difference between this pair of references lies in the UVLO levels like 16 V (on)/10 V (off) for the 42/44 versions and 8.4V (on)/7.6V (off) for the 43/45 types.

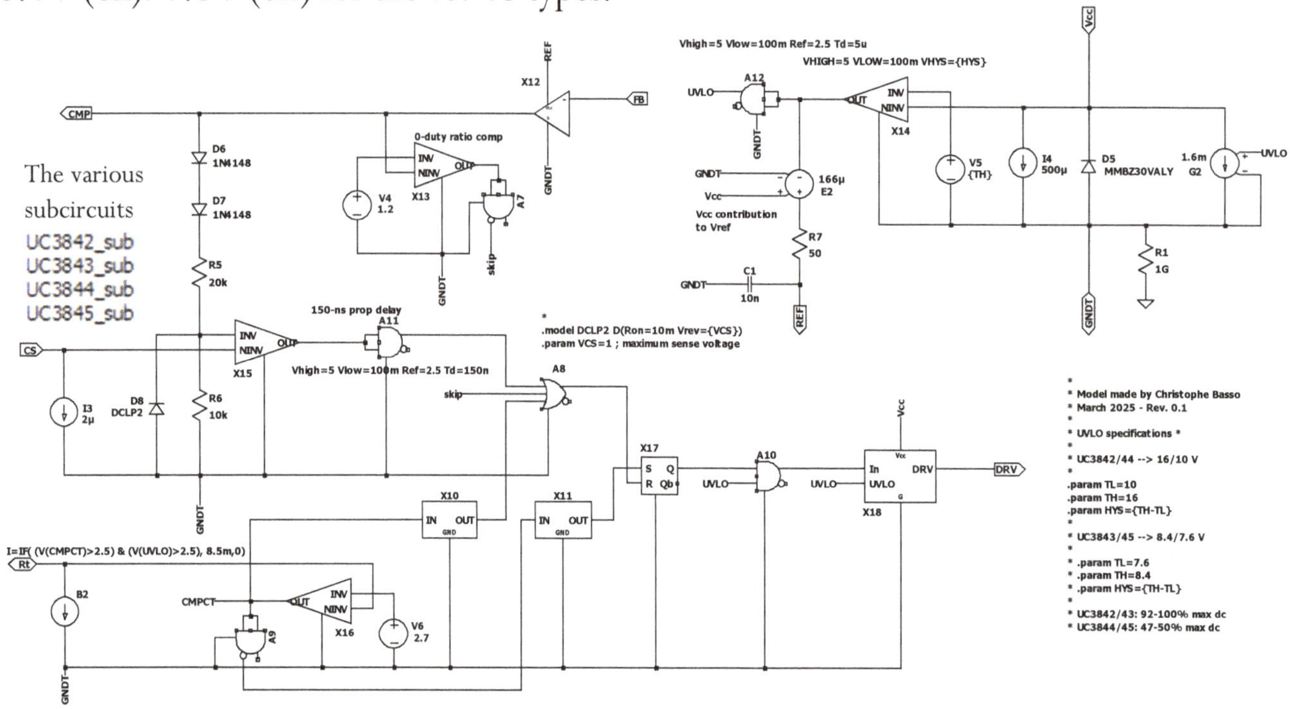

We start with the UVLO circuit built around comparator X_{14}. Depending on the hysteresis {HYS} and threshold values {TH}, it covers the four references. When node UVLO goes high, the circuit wakes up and draws a baseline of 1.6 mA. When supplemented with the drive pulses over a 1-nF capacitor, the averaged consumption amounts to 17 mA on average, as per the datasheet specs. You can freely change the start-up current (500 μA typical) and G_2. The error amplifier features two poles and reproduces the response of the datasheet. Gate A_{11} introduces a 150-ns propagation delay for modeling current overshoot. The reference voltage accounts for the V_{cc} line rejection via E_2. Please note that the DRV block X_{18} reproduces the real bipolar implementation. Finally, an extra flip-flop is added in the 50% duty-ratio versions.

Operating Parameters

BEFORE CLOSING the chapter on the generic controllers, it is important to explain what the parameters passed to these subcircuits are. Many are common to different structures such as those pertaining to the op-amp, while some are very specific like for an LLC controller, as an example.

- F_{sw} is the switching frequency
- Dutymax sets the maximum allowable duty ratio
- DT sets the dead time for a synchronous operation or a half-bridge driver
- V_{DRV} is the high-state output voltage of the driver. The low level is usually 100 mV
- I_{max} is the maximum current limit in volts observed at the CS pin (e.g. 1 V over 0.5 Ω gives 2 A as a maximum peak limit)
- A_{OL} lets you specify the open-loop gain of the internal op-amp (90 dB by default)
- Pole is the low-frequency pole (30 Hz by default) of the op-amp.
- I_{sink} and I_{source} program the maximum current capability of the error amplifier, with respectively 20 mA and 500 µA by default
- V_{OL} and V_{OH} program the allowable output voltage swing of the op-amp, respectively 100 mV and 3 V.
- V_v and V_p set the valley and peak voltage of the sawtooth for VM controllers. 0 V and 2 V are the default values.
- V_{ref} is the internal reference voltage, 2.5 V by default
- T_{ss} adjusts the soft-start time and is often set to 1 ms.
- R_{pullup} is the internal pull-up resistance when there is no op-amp in the subcircuit.
- F_{swmin} and F_{swmax} speak for themselves in the LLC controller
- L_{sw} for the *average* models (VM and CM) is the inductance of the converter, sometimes reflected back to the primary as in a forward converter.
- RISense specifies the sense resistance (sometimes designated as R_i) in the averaged CM PWM switch model.
- The slope {Se} corresponds to the stabilizing artificial ramp applied to the CM PWM switch model. It is expressed in volt per second.
- Other specific parameters will be described as they appear in the examples.

A Type 1 Compensator

THE TYPE 1 compensator is a simple integrator and provides no phase boost. It can be assembled around an op-amp, an operational transconductance amplifier (OTA) or a simple TL431 when used in conjunction with an optocoupler. In the collection of circuits, I have included many designs whose loop compensation is fully automated by a macro. However, it is also interesting to test the compensator separately and verify that the expected response matches the need.

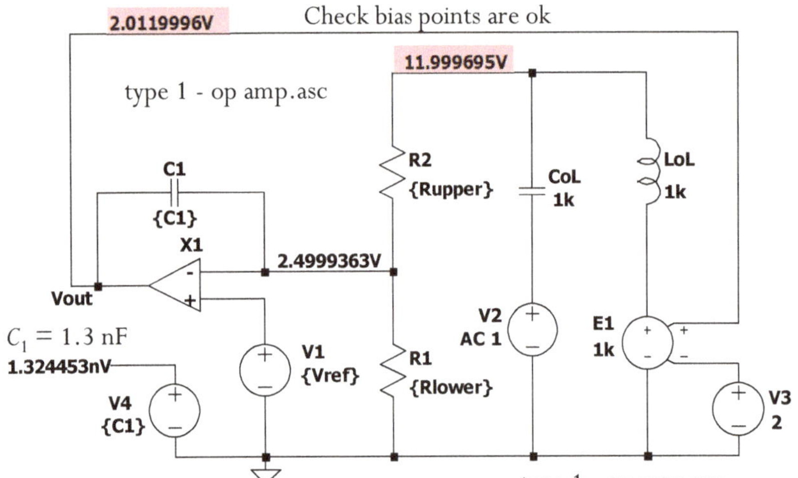

type 1 - op amp.asc

```
.param Gfc=-10 ; magnitude at crossover
*
.param fc=1k ; targeted crossover
*
.param Vout=12 ; output voltage
.param Ibias=250u ; bridge current
.param Vref=2.5
*
.param Rlower={Vref/Ibias}
.param Rupper={(Vout-Vref)/Ibias}
.param G={10**(-Gfc/20)}
.param Fpo={G*fc}
.param C1={1/(2*pi*Fpo*Rupper)}
*
```

You start with the control-to-output transfer function of the power stage (a boost, buck and so on) and extract the magnitude at f_c. Here, there is no phase boost, so we assume that only a gain or attenuation is needed to close the loop. In this example, we have a gain deficit of 10 dB at 1 kHz and the macro will determine C_1 to obtain a 10-dB gain at 1 kHz.

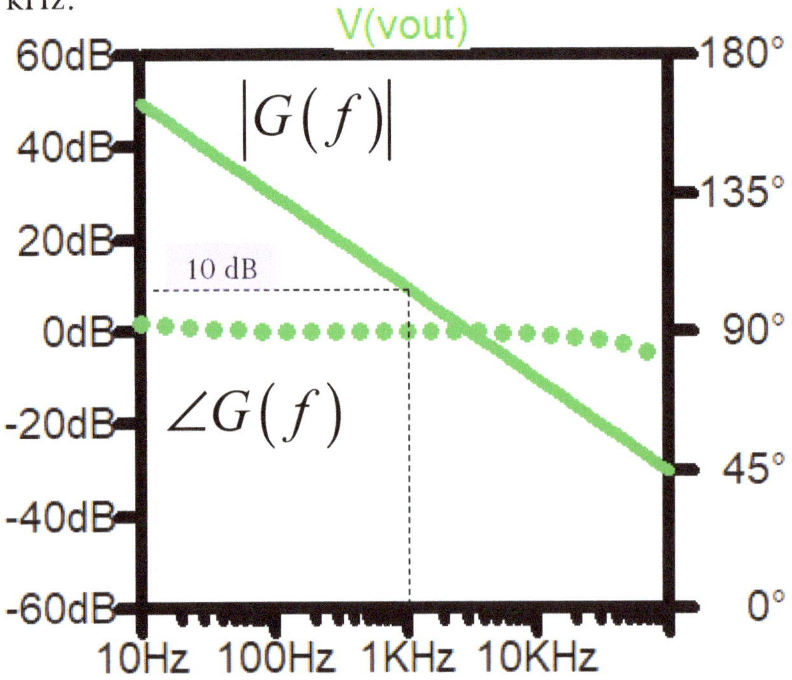

The magnitude drops with a -1-slope while the phase remains at 90°: the pole at the origin introduces a permanent phase lag of 90°, which, added to the 180° lag of the inverting op-amp, results in a phase of -270° or 90°. As expected, the gain is 10 dB at 1 kHz (a bit less considering the imperfect op-amp). The type 1 can be used in cases where the power stage exhibits a phase lag below 45°.

A Type 2 Compensator

THE TYPE 2 compensator keeps a pole at the origin (for a high dc gain, $s = 0$) but adds a pole-zero pair to boost the phase between 0 and 90°. If the pole and the zero are coincident, they neutralize each other and the response is flat in magnitude and phase. If you now place the zero before the pole, it will lead the phase up to 90°. When the pole kicks in, at a higher frequency, it will start to lag the phase, bringing the phase provided by the zero back to 0°. By adjusting the distance between the zero and the pole, with a crossover f_c usually selected in the geometric means between the two, you can adjust the boost as you wish. Keep in mind that the op-amp brings its own characteristics, like the GBW. You can therefore start the compensation exercise with my generic op-amp, but if you become ambitious in terms of crossover frequency (say beyond 10 kHz), check if the selected circuit can perform as expected and does not impose its own response over the expected type 2 you want.

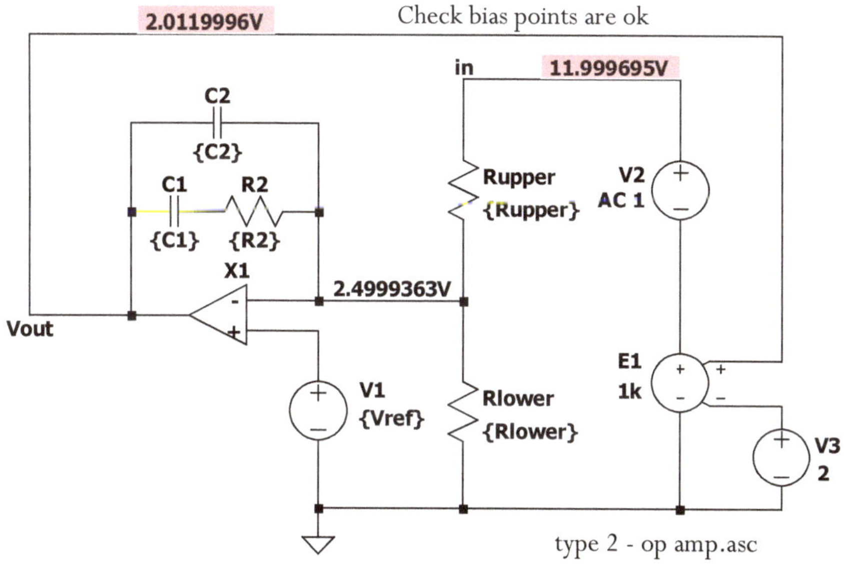


```
*
.param Gfc=-10 ; magnitude at crossover
.param PS=-90 ; phase lag at crossover
*
.param fc=1k ; targeted crossover
.param PM=60 ; choose phase margin
.param Vout=12
.param Ibias=250u
.param Vref=2.5
*
.param Rlower={Vref/Ibias}
.param Rupper={(Vout-Vref)/Ibias}
.param boost={PM-PS-90}
.param G={10**(-Gfc/20)}
.param kf={tan((boost/2+45)*pi/180)}
.param fp={fc*kf}
.param fz={fc/kf}
.param C2=1/{(2*pi*fc*G*kf*Rupper)}
.param C1={C2*(kf**2-1)}
.param R2={kf/(C1*2*pi*fc)}
*
```

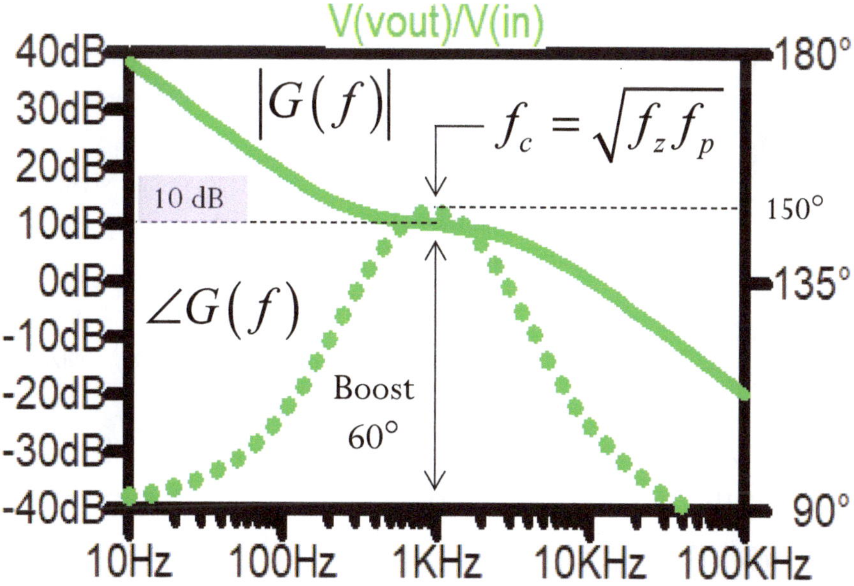

You can clearly see the phase bump of 60°, corresponding to our design goal (60° margin with 90° lag from the power stage). The peak lies in the geometric means of the pole and the zero (f_z = 268 Hz, f_p = 3.73 kHz).

A Type 3 Compensator

THE TYPE 3 compensator lets you theoretically boost the phase up to 180° via another pole-zero pair added to the previous type 2. We now have a pole at the origin – common to the three types – and a double pole-zero pair. If I used the classic *k* factor technique for the type 2, I prefer resorting to a manual pole-zero placement in the type 3. In this mode, I can freely place these inflexion points as I like, whereas the pole-zero pairs are coincident by nature in the *k* factor approach.

In the example, I have placed one zero at 1 kHz, the second at 2.5 kHz then a pole at half the switching frequency. The second pole is adjusted to meet the phase margin goal.

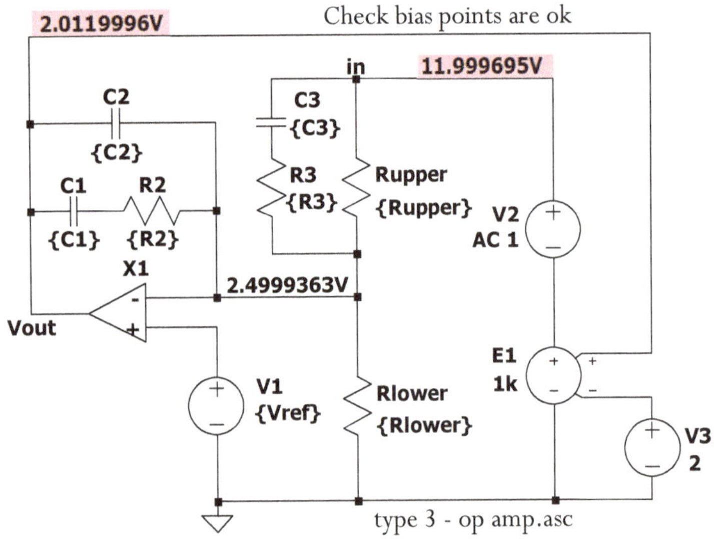

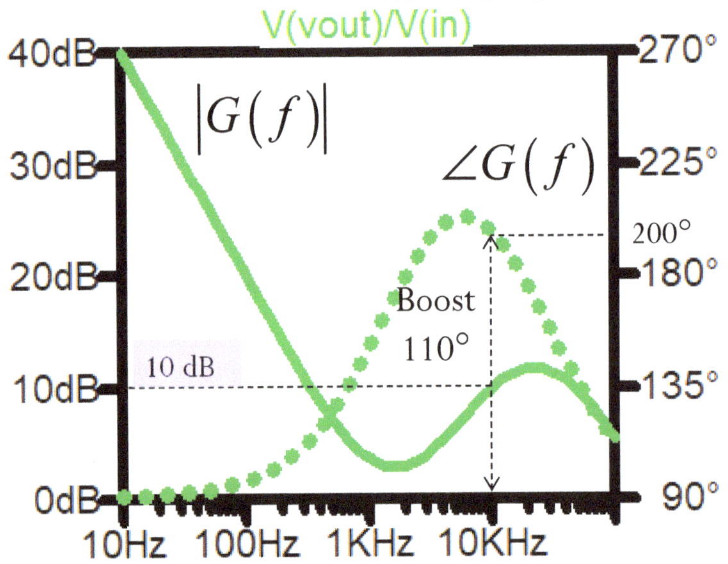


```
*
.param Gfc=-10 ; magnitude at crossover
.param PS=-130 ; phase lag at crossover
*
.param fc=10k ; targeted crossover
.param PM=70 ; choose phase margin
*
.param fz1=1k
.param fz2=2.5k
.param fp2=50k ; this pole placed at Fsw/2
*
.param Vout=12
.param Ibias=2m
.param Vref=2.5
*
.param Rlower={Vref/Ibias}
.param Rupper={(Vout-Vref)/Ibias}
.param boost={PM-PS-90}
.param G={10**(-Gfc/20)}
.param fp1={fc/tan(atan(fc/fz1)+atan(fc/fz2)-
atan(fc/fp2)-boost*pi/180)}
.param a={sqrt((fc**2/fp1**2)+1)}
.param b={sqrt((fc**2/fp2**2)+1)}
.param c={sqrt((fz1**2/fc**2)+1)}
.param d={sqrt((fc**2/fz2**2)+1)}
.param R2={((a*b/(c*d))/(fp1-fz1))*Rupper*G*fp1}
.param C1={1/(2*pi*fz1*R2)}
.param C2={C1/(C1*R2*2*pi*fp1-1)}
.param C3={(fp2-fz2)/(2*pi*Rupper*fp2*fz2)}
.param R3={Rupper*fz2/(fp2-fz2)}
.param G0={((R2*C1)/(Rupper*(C1+C2)))*c*d/(a*b)}
*
```

If f_{p1} is negative, zeros are not low enough to meet the phase boost target.

You can see how the phase bump is no longer peaking at f_c – it would with the *k* factor – and this is because the two poles and the two zeros are not coincident. In this example, pole f_{p1} is placed at 12.4 kHz while f_{p2} is located at 50 kHz.

Type 1 with an OTA

THE OTA LENDS itself well for building compensators. Integrated circuits designers like the OTA structure, because its die area is smaller than for an op-amp. An OTA is a voltage-controlled current source. There is no local feedback as with an op-amp and, therefore, the resistive divider sensing the output voltage now influences the transfer function. Also, the transconductance g_m expressed in siemens [S] plays a role. I recommend that you carefully look at its variability on the final compensator response. OTAs are popular in power factor correction (PFC) controllers. One reason is the ease of sensing an image of V_{out} at the FB pin, because there is no virtual ground as with an op-amp: detecting an over-voltage is thus easy. The second one is the ability to implement a so-called transient response enhancer which quickly charges capacitor C_1 in case the loop is slow as in PFCs.

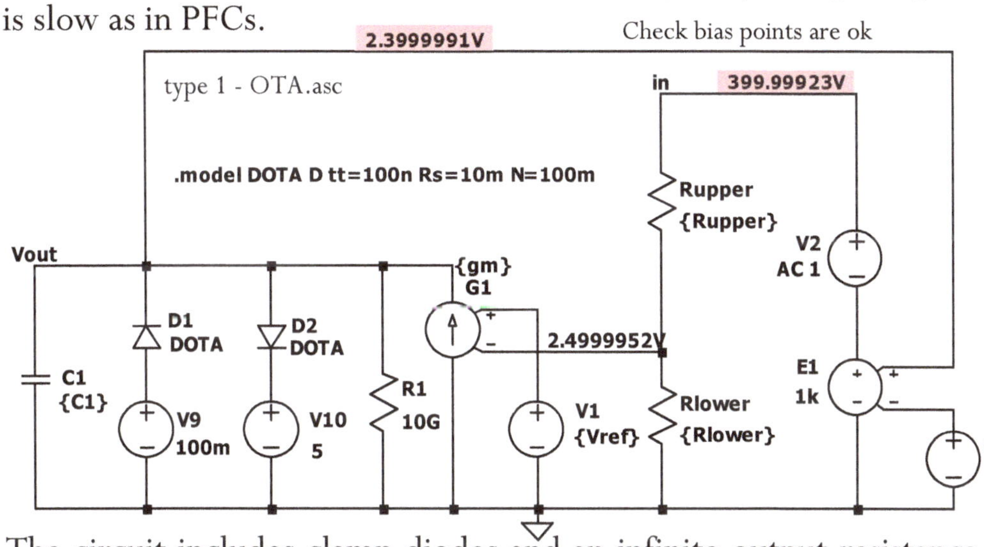

The OTA output network (output resistance r_o, ESD structure) plays a role in the ac response and this paper [4] is an interesting read to understand their impact on the design.

The circuit includes clamp diodes and an infinite output resistance R_1. The latter plays a role in the frequency response – the open-loop gain for instance – and you change its value if need be. The bias points show that the output of E_1 is well at 400 V, as expected for this PFC (f_c = 20 Hz).

```
*
.param Gfc=20 ; magnitude at crossover
.param fc=20 ; targeted crossover
*
.param Vout=400
.param Ibias=250u
.param Vref=2.5
*
.param gm=50u ; transconductance in Siemens
.param Rlower={Vref/Ibias}
.param Rupper={(Vout-Vref)/Ibias}
.param G={10**(-Gfc/20)}
.param fp0={fc*G}
.param C1={gm*Rlower/(2*pi*(Rlower+Rupper)*fp0)}
*
```

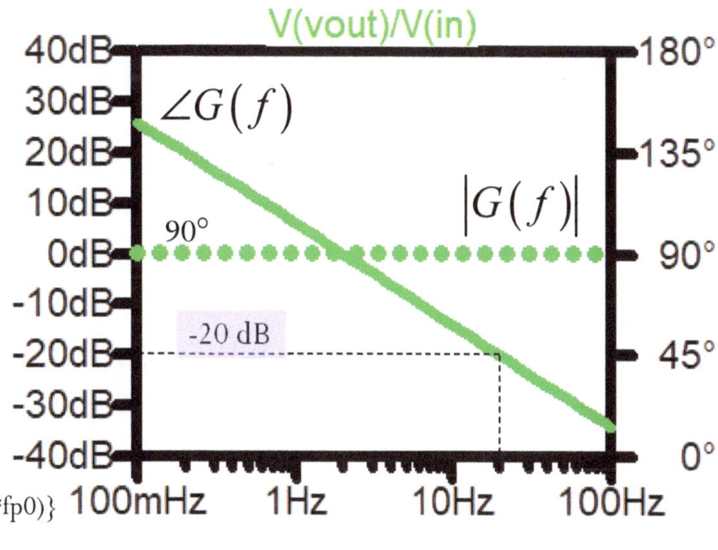

Type 2 with an OTA

A TYPE 2 CONFIGURATION is obtained by adding an extra *RC* network across the previous integration capacitor. The type 2 is found in a vast majority of converters, including PFCs where some phase boost is needed to avoid a too large output overshoot.

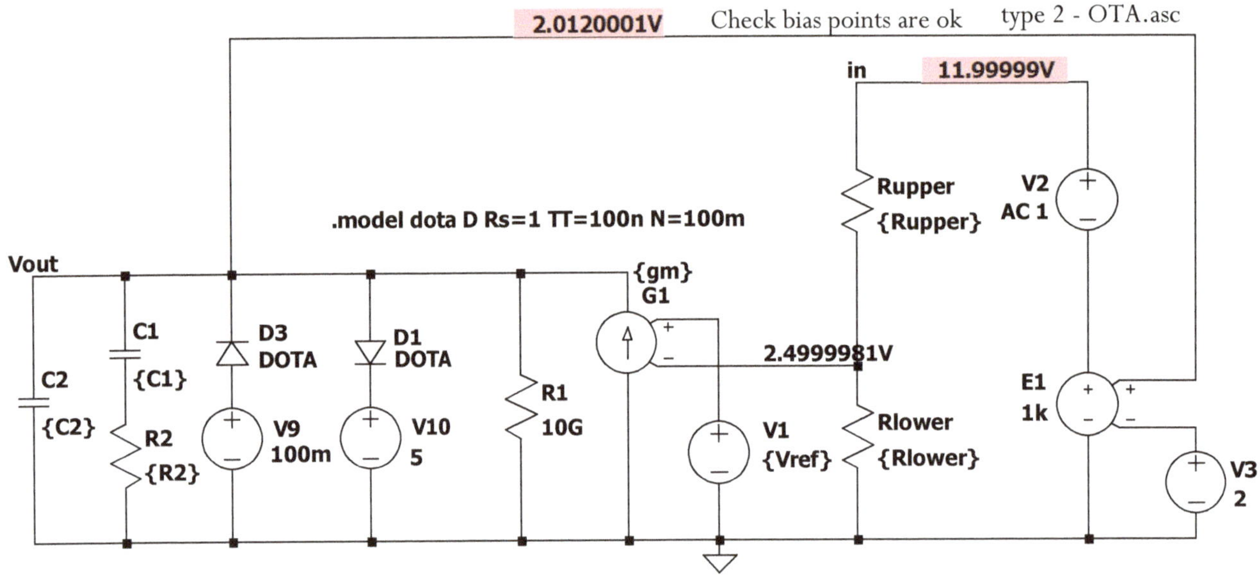

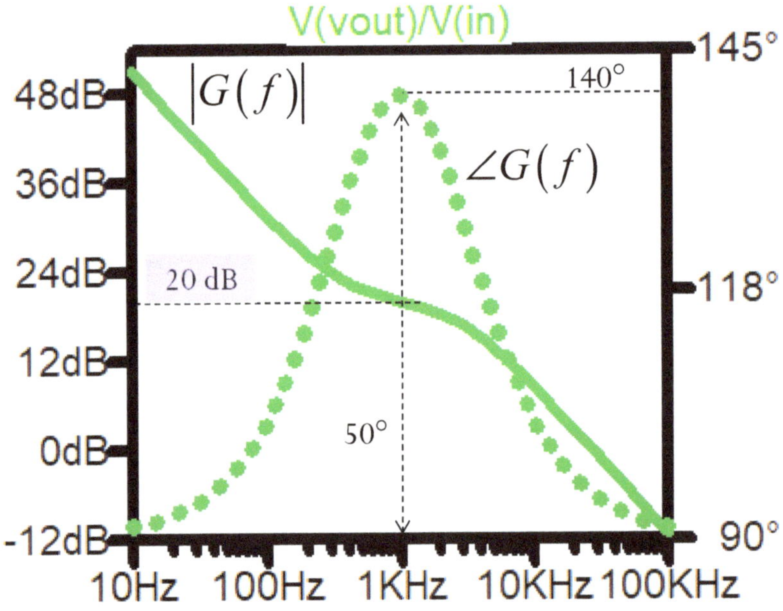

I will not present a type 3 around an OTA because it is extremely impractical to design: the distance between the pole-zero of the second pair, depends on the ratio between the regulated voltage and the reference value. If you regulate a 5-V rail from a 2.5-V reference, it is almost impossible to obtain the maximum phase boost an op-amp would give. Check my APEC 2010 seminar for more details.

```
*
.param Gfc=-20 ; magnitude at crossover
.param PS=-70 ; phase lag at crossover
*
.param fc=1k ; targeted crossover
.param PM=70 ; choose phase margin
*
.param Vout=12
.param Ibias=250u
.param Vref=2.5
*
.param gm=100u ; transconductance
.param Rlower={Vref/Ibias}
.param Rupper={(Vout-Vref)/Ibias}
.param boost={PM-PS-90}
.param G={10**(-Gfc/20)}
.param kf={tan((boost/2+45)*pi/180)}
.param fp={fc*kf}
.param fz={fc/kf}
.param a={sqrt((fc**2/fp**2)+1)}
.param b={sqrt((fz**2/fc**2)+1)}
.param R2={(a/b)*(fp*G)*(Rlower+Rupper)/
((fp-fz)*Rlower*gm)}
.param C1={1/(2*pi*R2*fz)}
.param C2={Rlower*gm*(b/a)/
(2*pi*fp*G*(Rlower+Rupper))}
*
```

A Type 1 with the TL431

A SIMPLE INTEGRATOR can also be built around a TL431 and an optocoupler. The macro starts by calculating the resistive divider made of R_{upper} and R_{lower} for regulating at 12 V in this example. Then, depending on the pull-up resistance and the internal V_{dd}, the maximum LED resistance is determined. It accounts for an extra bias current forced in the LED, to fulfil the minimum 1-mA bias current of the TL431. It is usually set by a resistance placed across the optocoupler LED, R_{bias}. With a typical value of 1 kΩ across a ≈1-V drop, an extra 1-mA current is created at no expense. This current limits the maximum resistance for R_{LED} which, if selected too high, can potentially bring a loss of regulation. In the macro I propose, I select 50% of the maximum series LED resistance. This extra resistance, paralleled with R_{LED}, can be omitted for a TLV431 which operates down to a few hundreds of micro-amperes. Finally, this extra resistance diverts a little bit of feedback current and reduces the final gain, especially if you operate the LED at a low bias current, where its dynamic resistance r_d can be rather high and no longer be neglected.

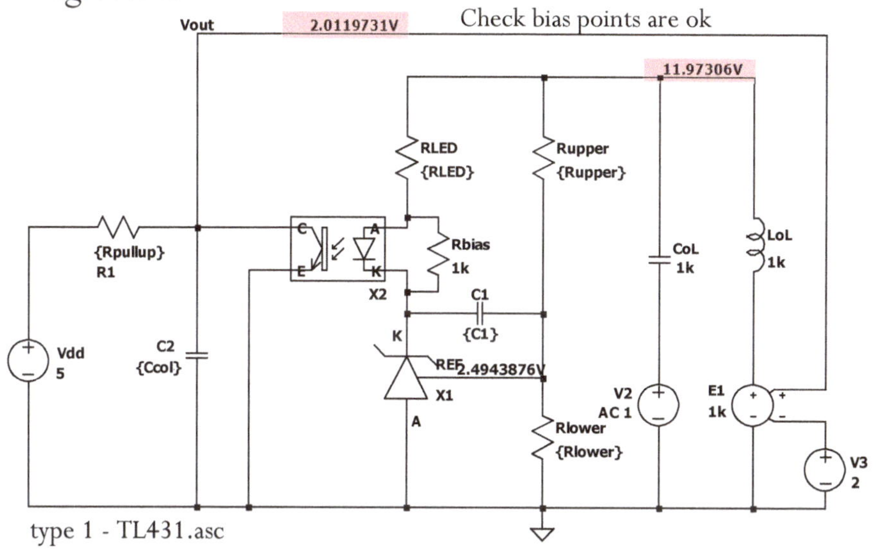

type 1 - TL431.asc

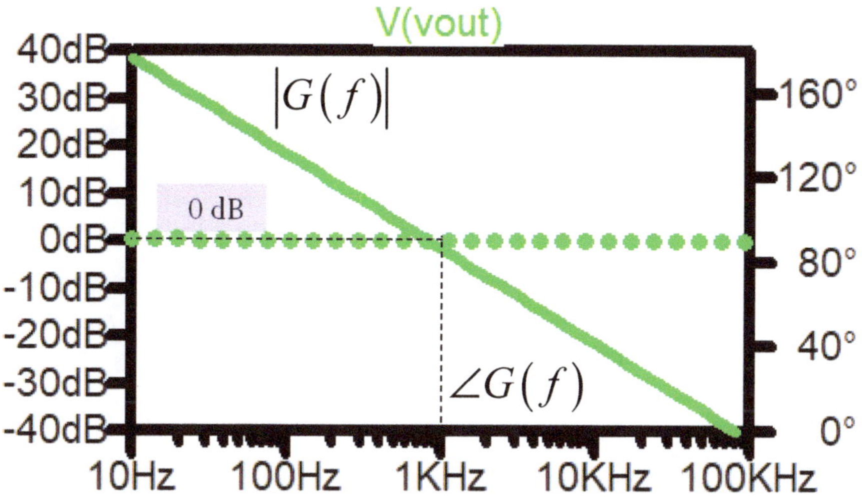


```
*
.param Gfc=0 ; magnitude at crossover
.param fc=1k ; targeted crossover
*
.param Vout=12
.param Ibias=250u
.param Vref=2.5
.param Rpullup=10k
*
.param Fopto=15k
.param Copto={1/(2*pi*Fopto*Rpullup)}
.param CTR=1
*
.param VL=2.5 ; min TL431 voltage
.param VCEsat=0.3 ; Opto sat voltage
.param Vdd=5 ; pull-up resistance bias
.param Vf=1 ; LED forward voltage
.param Ib=1m ; 1k resistance across the LED
.param A={Vout-Vf-VL}
.param B={Vdd-VCEsat+Ib*CTR*Rpullup}
.param Rmax={(A/B)*Rpullup*CTR}
.param RLED={0.5*Rmax}
*
.param Rlower={Vref/Ibias}
.param Rupper={(Vout-Vref)/Ibias}
.param G={10**(-Gfc/20)}
.param Fpo={G*fc}
.param C2={CTR/(2*pi*Fpo*RLED)}
.param C1={C2*Rpullup/Rupper}
.param Ccol={C2-Copto}
*
```

A Type 2 with the TL431

THE TYPE 2 compensator with a TL431 is probably the most popular structure found in ac-dc adapters for the consumer market. Its implementation is similar to that of the previous type 1 example, except that the pole and zero are no longer coincident to create phase boost. Please note that a *single* capacitor is enough, not an *RC* network as I often see in design. Owing to the fast lane presence, a zero is naturally created with one capacitor.

Because of this configuration, there is a limit in the *minimum* gain this architecture can bring and this gain depends on R_{LED}. If R_{LED} is selected too high, in an attempt to reduce the mid-band gain, then you can lose regulation as the limit set by R_{max} in the macro may be violated. In the type 1, this was not a problem, as R_{max} could be easily adjusted as the final gain was set by the 0-dB crossover pole f_{po}. But not here, so you must check this point, especially if you want to regulate with a low output voltage, like 5 V or below. You may need to toggle to a TLV431, as its minimum operating voltage is 1.24 V versus 2.5 V for a regular TL431. Another option will be to disable the fast lane.

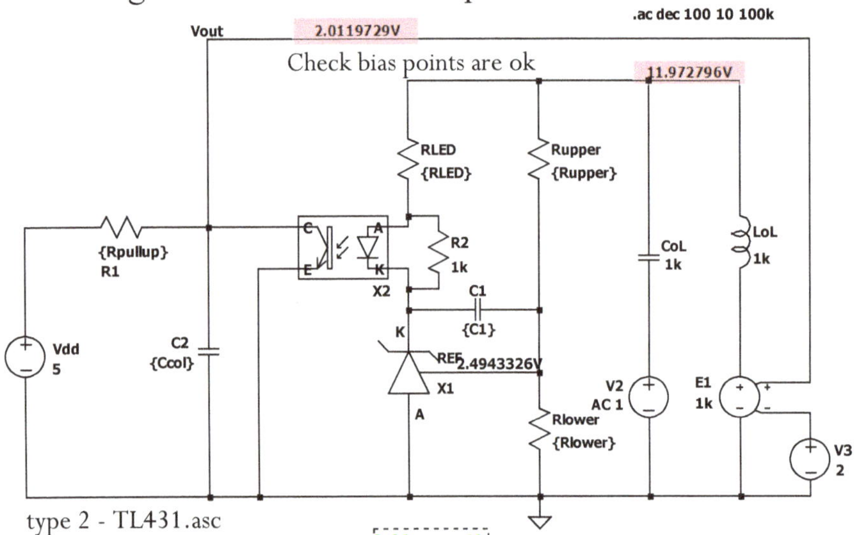

type 2 - TL431.asc

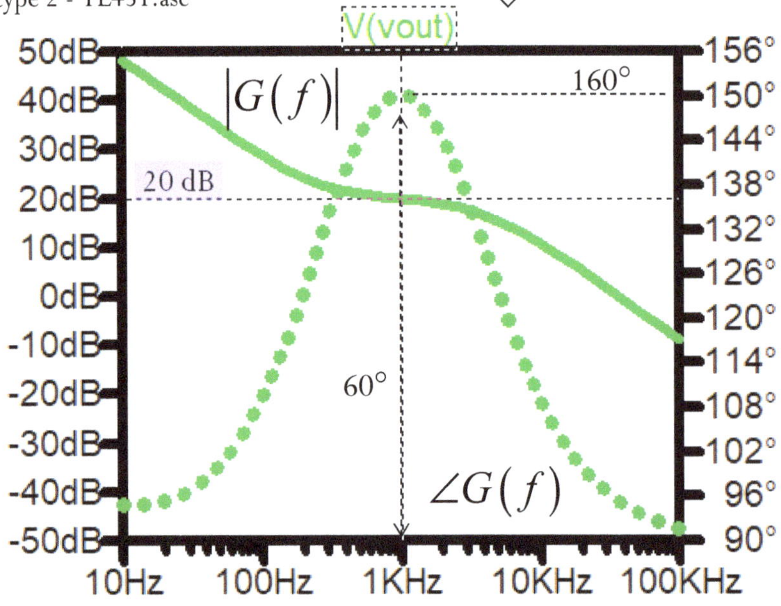


```
.param Gfc=-20 ; magnitude at crossover
.param PS=-90 ; phase lag at crossover
*
.param fc=1k ; targeted crossover
.param PM=60 ; choose phase margin
*
.param Vout=12
.param Ibias=250u
.param Vref=2.5
.param Rpullup=20k
*
.param Fopto=6k
.param Copto={1/(2*pi*Fopto*Rpullup)}
.param CTR=0.35
*
.param VL=2.5 ; min TL431 voltage
.param VCEsat=0.3 ; Opto sat voltage
.param Vdd=5 ; pull-up resistance bias
.param Vf=1 ; LED forward voltage
.param Ib={Vf/1k} ; 1k paralleled with LED
.param A={Vout-Vf-VL}
.param B={Vdd-VCEsat+Ib*CTR*Rpullup}
.param Rmax={(A/B)*Rpullup*CTR}
.param Rlower={Vref/Ibias}
.param Rupper={(Vout-Vref)/Ibias}
.param boost={PM-PS-90}
.param G={10**(-Gfc/20)}
.param kf={tan((boost/2+45)*pi/180)}
.param fp={fc*kf}
.param fz={fc/kf}
.param RLED={CTR*Rpullup/G}
.param C1={1/(2*pi*fz*Rupper)}
.param C2={1/(2*pi*fp*Rpullup)}
.param Ccol={C2-Copto}
```

A Type 2 with the TL431 – no Fast Lane

THE FAST LANE can be an obstacle in different situations:

1. V_{out} is too small to ensure enough voltage for biasing the LED and the TL431
2. V_{out} is too high, e.g. 48 V, and exceeds the TL431 breakdown voltage (37 V)
3. The mid-band gain you need is incompatible with the bias requirements set by the maximum R_{LED} value (R_{max} in the macro)

For the above reasons, you will disconnect the fast lane and bias the LED resistance with a fixed dc voltage, well decoupled from V_{out}. In the below example, I have chosen a 5-V V_{aux} and a LED resistance equal to 80% of the maximum calculated value. Capacitor C_3 is placed across the TL431 for noise reasons only, while the pole is set by C_2, in the primary side. The configuration brings a 10-dB attenuation at 1 kHz.

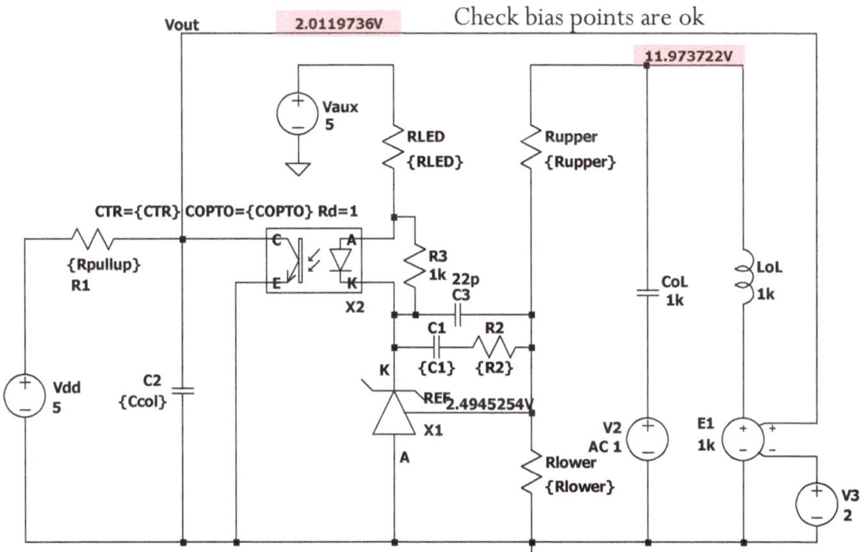

type 2 - TL431.asc

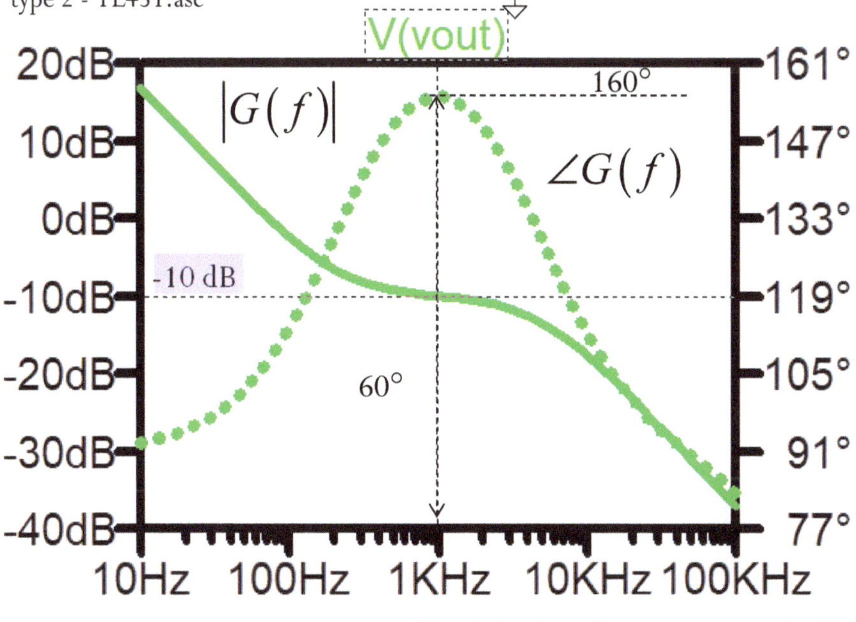


```
.param Gfc=10 ; magnitude at crossover
.param PS=-85 ; phase lag at crossover
*
.param fc=1k ; targeted crossover
.param PM=70 ; choose phase margin at fc
*
.param Vout=12
.param Ibias=250u
.param Vref=2.5
.param Rpullup=10k
.param Vaux=5
.param VL=2.5
.param VCEsat=0.3
.param Vdd=5
.param Ib=1m
.param Vf=1
.param A={Vaux-Vf-VL}
.param B={Vdd-VCEsat+Ib*CTR*Rpullup}
.param Rmax={(A/B)*Rpullup*CTR}
.param RLED={0.8*Rmax}
*
.param Rlower={Vref/Ibias}
.param Rupper={(Vout-Vref)/Ibias}
.param boost={PM-PS-90}
.param kf={tan((boost/2+45)*pi/180)}
.param G1={Rpullup*CTR/RLED}
.param G2={10**(-Gfc/20)}
.param G={G2/G1}
.param fp={fc*kf}
.param fz={fc/kf}
.param d={(fz**2+fc**2)*(fp**2+fc**2)}
.param c={(fz**2+fc**2)}
.param R2={(sqrt(d)/c)*G*fc*Rupper/fp}
.param C1={1/(2*pi*fz*R2)}
.param C2={1/(2*pi*fp*Rpullup)}
.param Ccol={C2-Copto}
```

A Type 3 with the TL431

THE TL431 does not really lend itself well for implementing a type 3 compensator. The guilty is the fast lane brought by R_{LED} sensing V_{out}, which, in this case, affects the gain, the second zero position and the bias limit. Therefore, if you plan on implementing this compensator, make sure you understand the limits and sweep the moving parameters like the optocoupler CTR or the internal pull-up resistance. Make sure regulation is never lost, with margins, when all these contributors move along production and the converter lifetime. If this is too marginal, disconnect the fast lane as explained in the next page.

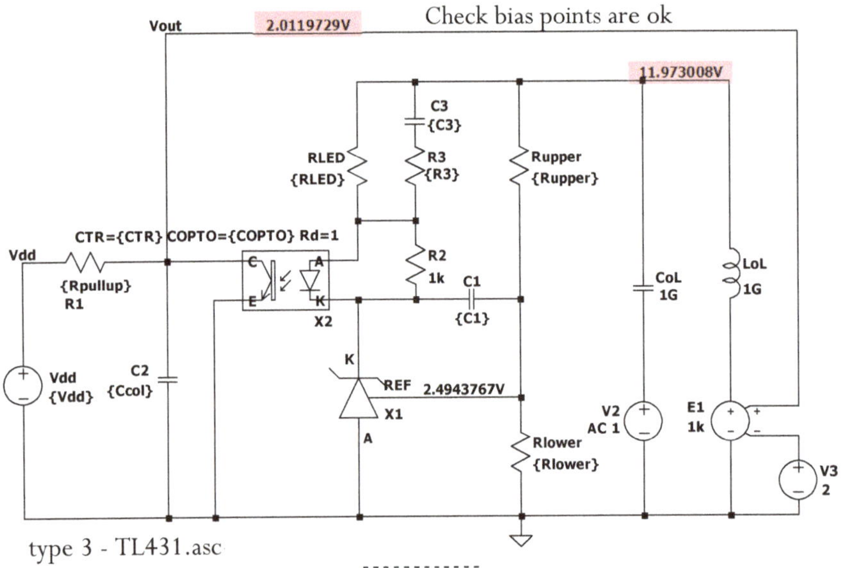

type 3 - TL431.asc

```
*
.param Gfc=-20 ; magnitude at crossover
.param PS=-140 ; phase lag at crossover
.param fc=2k ; targeted crossover
.param PM=70 ; choose phase margin
*
.param Vout=12
.param Ibias=250u
.param Vref=2.5
.param Rlower={Vref/Ibias}
.param Rupper={(Vout-Vref)/Ibias}
.param Rpullup=10k
*
.param Fopto=15k
.param Copto=1/(2*pi*Fopto*Rpullup)
.param CTR=0.85
*
.param boost=PM-PS-90
.param G=10**(-Gfc/20)
*
.param fz1=400
.param fz2=600
.param fp2=50k ; half Fsw
.param fp1={fc/tan(atan(fc/fz1)+atan(fc/fz2)-
atan(fc/fp2)-boost*pi/180)}
*
.param C1={1/(2*pi*fz1*Rupper)}
.param C3={((fp2-
fz2)/(fz2*fp2))*G*F/(Rpullup*CTR)/(2*pi)}
.param R3={((fz2/(fp2-fz2))*CTR*Rpullup/G)*E}
.param a={sqrt((fc**2/fp1**2)+1)}
.param b={sqrt((fc**2/fp2**2)+1)}
.param c={sqrt((fz1**2/fc**2)+1)}
.param d={sqrt((fc**2/fz2**2)+1)}
.param E={c*d/(a*b)}
.param F={a*b/(c*d)}
.param RLED={(CTR*Rpullup/G)*E}
.param C2={1/(2*pi*fp1*Rpullup)}
.param Ccol={C2-Copto}
*
```

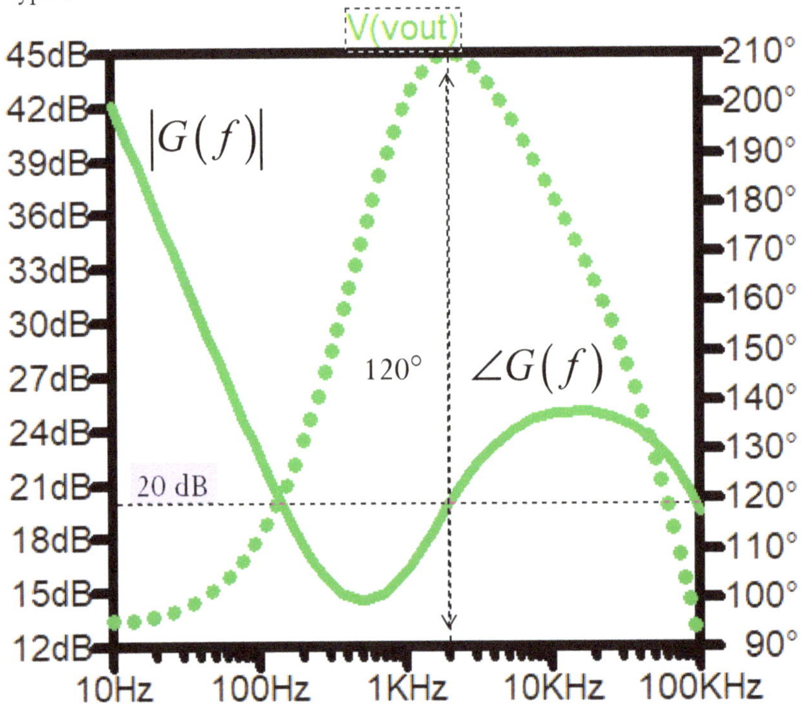

The phase boost is exactly 120° at 2 kHz, with an expected gain of 20 dB. The overall shape is due to the asymmetrical pole-zero pair placement.

A Type 3 with TL431 – no Fast Lane

THE FAST LANE issue originates in the LED current, which is supposedly imposed by the TL431 alone. At high frequencies, the ac path through the TL431 should be inhibited by capacitor C_1, which becomes a short circuit. However, even if no ac current flows in the TL431 in this mode, its cathode is always kept at the same potential for ensuring dc regulation. A voltage node fixed in dc is equivalent to zero volt in ac, implying that the LED anode is grounded in ac. This offers an extra direct path to V_{out} via R_{LED}: this is the fast-lane effect. We are able to take advantage of it in the type 2 configuration – hence only C_1 alone is necessary – but it becomes an obstacle to flexibility in the type 3. The solution? As in the type 2 example, disconnect the fast lane and hook the LED resistance to a 0-V in ac, e.g. a stable dc voltage, fully decoupled from V_{out}.

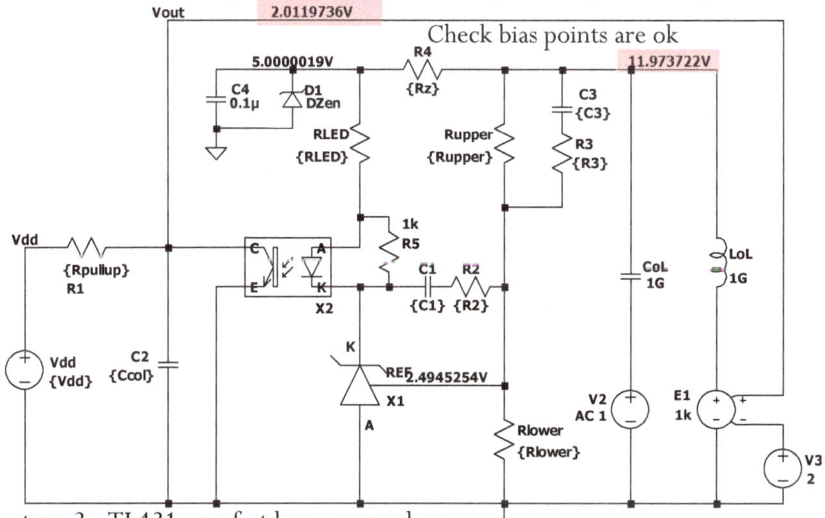

type 3 - TL431 - no fast lane - manual.asc

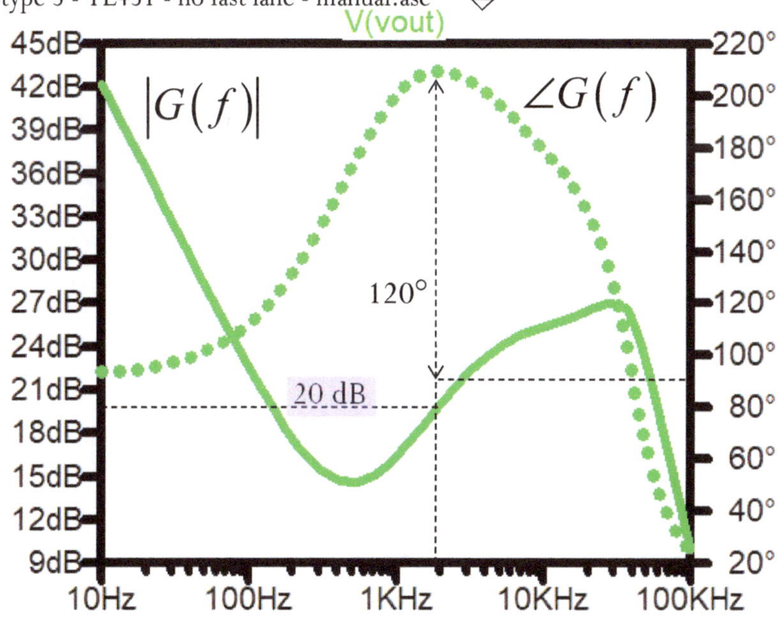

The fast-lane decoupling depends on the ratio between the selected Zener voltage and V_{out}. In lack of perfect rejection, the ac response is slightly distorted.

```
.param Gfc=-20 ; magnitude at crossover
.param PS=-140 ; phase lag at crossover
*
.param fc=2k ; targeted crossover
.param PM=70 ; choose phase margin at crossover
*
.param Vout=12
.param Ibias=250u
.param Vref=2.5
.param Rlower={Vref/Ibias}
.param Rupper={(Vout-Vref)/Ibias}
.param Rpullup=10k
.param Vz=5 ; selected Zener voltage
.param IZbias=2m ; Zener bias current
*
.param Fopto=15k
.param Copto=1/(2*pi*Fopto*Rpullup)
.param CTR=0.85
*
.param boost=PM-PS-90
.param G=10**(-Gfc/20)
*
.param fz1=400
.param fz2=600
.param fp2=50k ; half Fsw
.param fp1={fc/tan(atan(fc/fz1)+atan(fc/fz2)-
atan(fc/fp2)-boost*pi/180)}
.param G1={Rpullup*CTR/RLED}
.param G2={Rupper*G/G1}
.param a={sqrt((fc**2/fp1**2)+1)}
.param b={sqrt((fc**2/fp2**2)+1)}
.param c={sqrt((fz1**2/fc**2)+1)}
.param d={sqrt((fc**2/fz2**2)+1)}
.param R2={(a*b/(c*d))*G2}
.param C1={1/(2*pi*fz1*R2)}
.param C2={1/(2*pi*fp1*Rpullup)}
.param C3={(fp2-fz2)/(2*pi*Rupper*fp2*fz2)}
.param R3={Rupper*fz2/(fp2-fz2)}
.param Ccol={C2-Copto}
*
```

A Simple Filtered PID

THE PID COMPENSATOR is a popular circuit used to close the loop in a variety of applications. If most tuning methods involve a kind of trial-and-error process, I prefer converting the poles and zeros I need into PID coefficients, leading exactly to the expected response. A PID involves a pole at the origin with two zeros, and can be implemented in different manners. The *filtered* PID adds an extra pole to flatten out the response at high frequencies and improve the noise susceptibility of the loop. The example below is a classic in a control theory course. The macro determines the values for the *RC* components around a perfect voltage-controlled source. I then did add Laplace expressions to compare the various formulations of the filtered PID compensators.

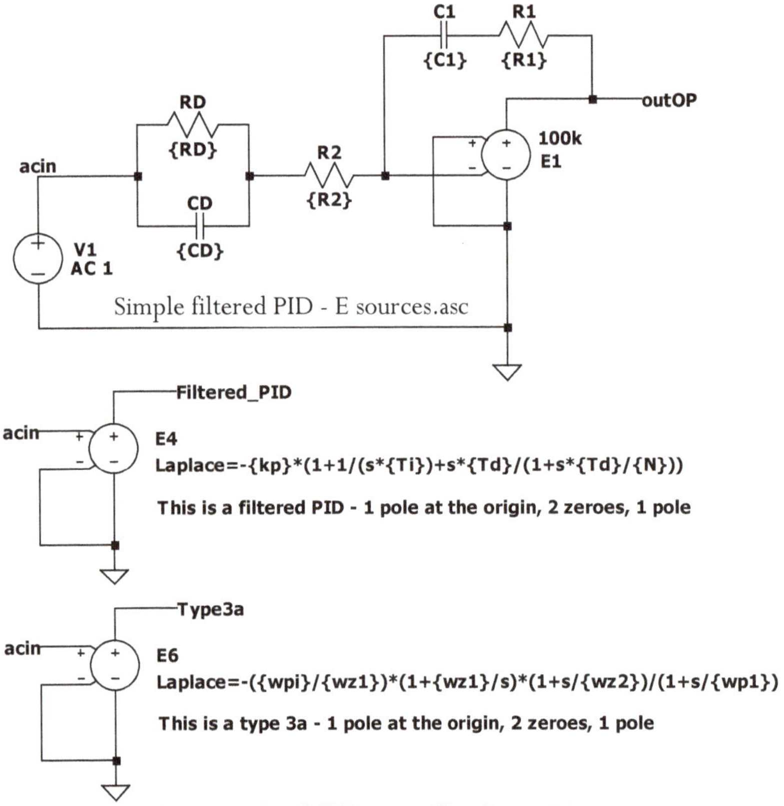


```
.param Gfc=10 ; gain (in dB) you want a crossover
.param fc=1k ; targeted crossover
.param G=(10**(Gfc/20))
*
.param R1=22k ; R1 is arbitrarily selected
.param fz1=200
.param fz2=600
.param fp1=6k ; this is the PID pole in the filtered version
*
.param wz1={2*pi*fz1}
.param wz2={2*pi*fz2}
.param wp1={2*pi*fp1}
*
.param j={(1+fz1**2/fc**2)*(1+fc**2/fz2**2)}
.param G0={G*sqrt((1+fc**2/fp1**2)/j)}
.param Wpi={wz1*G0}
.param z={(Wp1-Wz1)*(Wp1-Wz2)}
.param f={Wp1*Wz1+Wp1*Wz2-Wz1*Wz2}
*
.param Td={z/(f*Wp1)}
.param N={((Wp1**2)/f)-1}
.param kp={(Wpi/Wz1)-(Wpi/Wp1)+(Wpi/Wz2)}
.param Ti={((Wz1+Wz2)/(Wz1*Wz2))-(1/Wp1)}
.param ki={kp/Ti}
.param kd={kp*Td}
*
.param A={sqrt(N**2*Ti**2-4*N**2*Td*Ti-2*N*Td*Ti+Td**2)}
.param C1={(Td+A+N*Ti)/(2*N*R1)}
.param CD={kp*Td*(1+N)*(2*Ti*N**2+Ti*N+A-Td)/(2*N**3*R1*Ti)}
.param R2={R1/(kp+N*kp)}
.param RD={R1*(N*Ti-Td-A)/(2*kp*Td*(1+N))}
```

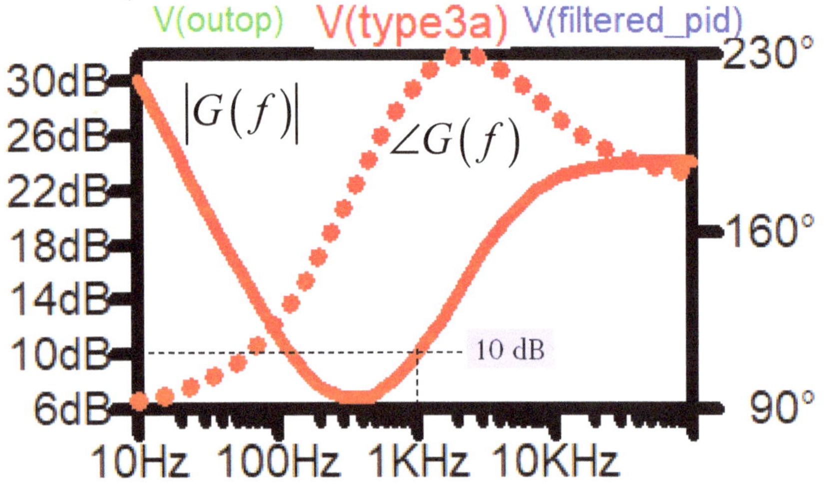

The ac responses of the op-amp-based filtered PID and the Laplace expressions, are identical.

PID to Type 3 Compensator

A PID IS RARELY used in its simplest form with switching converters. As explained in the previous example, a pole is added to the original structure, to form a *filtered* PID, forcing the gain to roll off at high frequency. If another pole is added to the filtered PID, you have a type 3 compensator, well known in our field. In this example, I determine the coefficients by placing poles and zeros in relationship with the gain and phase margin I want. The PID terms k_p, k_i, k_d with the two extra poles are computed by the macro.

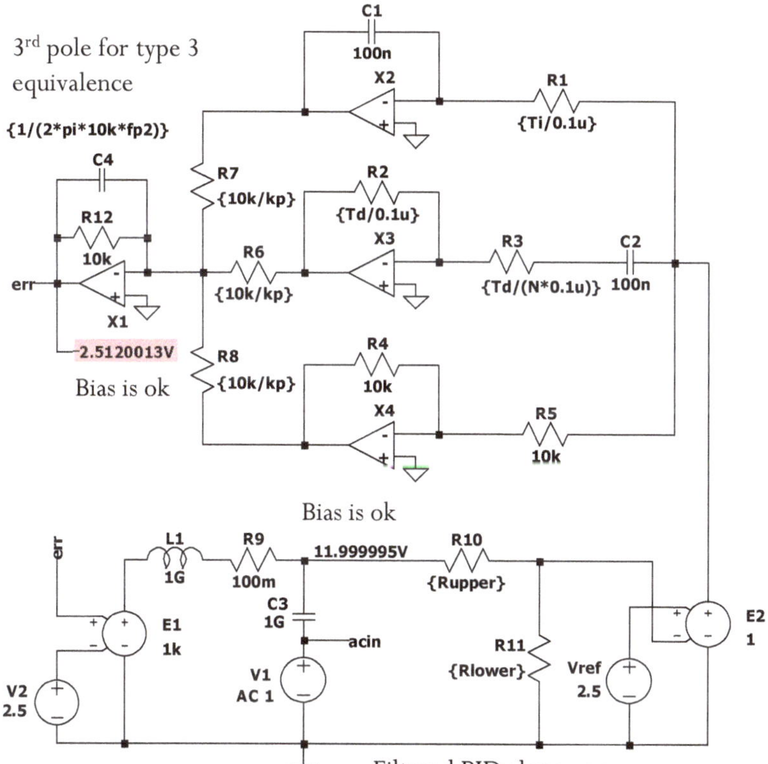

Filtered PID alone - op amps.asc

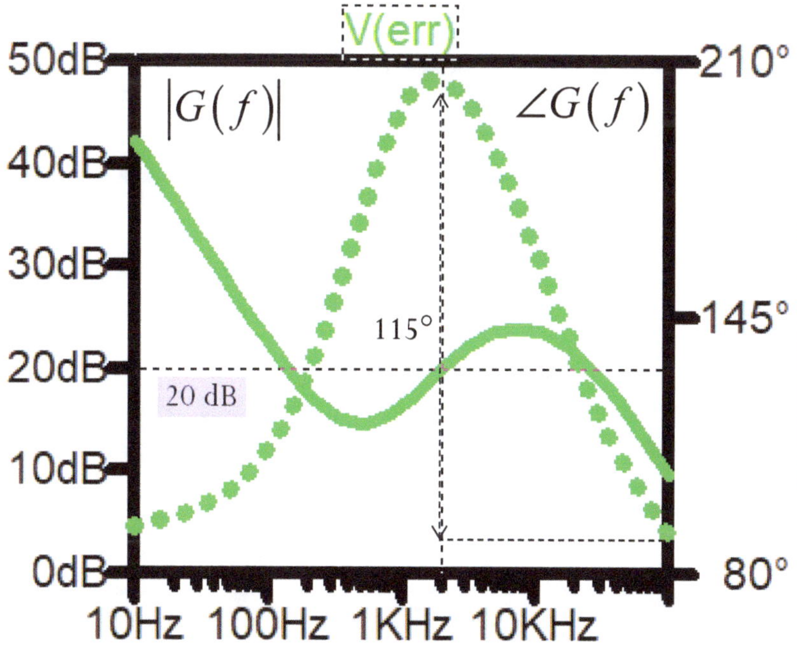


```
.param Gfc=-20 ; magnitude at crossover
.param PS=-140 ; phase lag at crossover
.param fc=2k ; targeted crossover
.param PM=70 ; choose phase margin at crossover
*
.param Vout=12
.param Ibias=1m
.param Vref=2.5
.param Rlower=Vref/Ibias
.param Rupper=(Vout-Vref)/Ibias
.param kdiv=Rlower/(Rlower+Rupper)
*
.param fz1=400
.param fz2=600
.param fp2=50k
*
.param boost=PM-PS-90
.param G=(10**(-Gfc/20))/kdiv
.param fp1=fc/tan((atan(fc/fz1)+atan(fc/fz2)-
atan(fc/fp2))-boost*pi/180)
*
.param wz1={2*pi*fz1}
.param wz2={2*pi*fz2}
.param wp1={2*pi*fp1}
.param wp2={2*pi*fp2}
*
.param i={(1+fc**2/fp1**2)*(1+fc**2/fp2**2)}
.param j={(1+fz1**2/fc**2)*(1+fc**2/fz2**2)}
.param Wpi={sqrt(i/j)*G*fz1*2*pi}
.param z={(Wp1-Wz1)*(Wp1-Wz2)}
.param f={Wp1*Wz1+Wp1*Wz2-Wz1*Wz2}
*
                                        PID terms
.param Td={z/(f*Wp1)}
.param N={((Wp1**2)/f)-1}
.param kp={(Wpi/Wz1)-
(Wpi/Wp1)+(Wpi/Wz2)}
.param Ti={((Wz1+Wz2)/(Wz1*Wz2))-(1/Wp1)}
.param ki={kp/Ti}
.param kd={kp*Td}
*
```

The ac response is that of a type 3 compensator with all the PID terms automated by the macro.

A Digital Type 2 Compensator

WITHOUT ENTERING into the details of coding digital compensators, it is always interesting to understand how filtering is implemented in the discrete world, and check what the ac response would be, once the coefficients have been determined. This is the object of the following pages which describe the most popular compensators, types 2 and 3. We start with a type 2, implemented as a biquad filter.

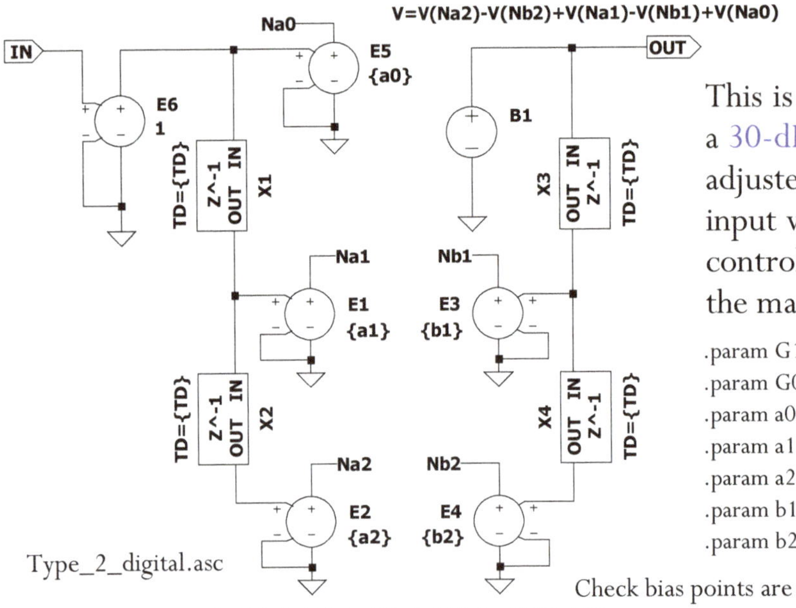

This is an example for a 400-V PFC with a 30-dB excess of gain. This gain could be adjusted on the fly, depending on the input voltage, a benefit brought by digital control. All coefficients are automated by the macro.

```
.param G1={10**(-Gfc/20)*(Rlower+Rupper)/Rlower}
.param G0={IF(Vin<150,G1,G1/4)}
.param a0={G0*Tsw*wp*(Tsw*wz+2)/(2*Tsw*wp+4)}
.param a1={G0*Tsw**2*wp*wz/(Tsw*wp+2)}
.param a2={G0*Tsw*wp*(Tsw*wz-2)/(2*Tsw*wp+4)}
.param b1={-8/(4+2*Tsw*wp)}
.param b2={(4/(Tsw*wp+2))-1}
```

Check bias points are ok

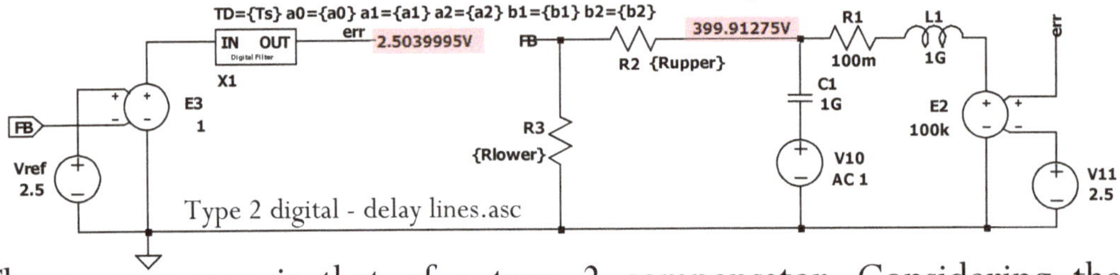

The circuit requires an auto-bias generator to obtain the right operating point.

The ac response is that of a type 2 compensator. Considering the wanted 10-Hz crossover frequency and the 10-µs sampling period, the response is identical to that of an analog version. There is no zero-order hold (ZOH) in this implementation.

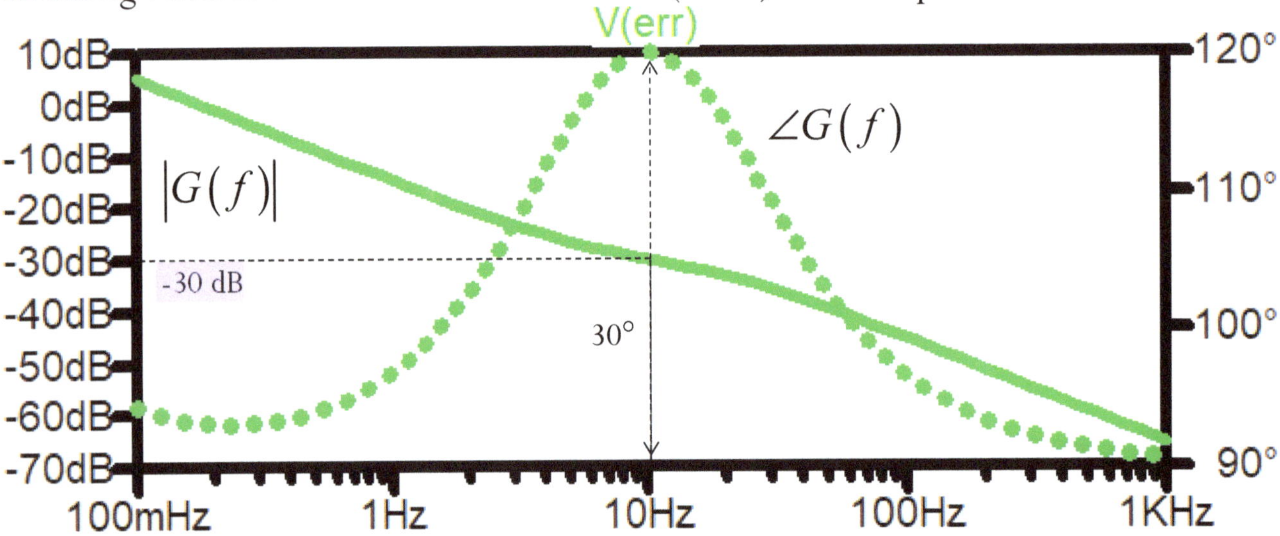

A Digital Type 3 Compensator - I

A TYPE 3 COMPENSATOR is not more complicated that its type 2 counterpart. The procedure remains similar: start from the *low-entropy* analog transfer function and replace s using the bilinear transform. You obtain a polynomial for the numerator and the denominator. Factor them in a convenient way by dividing both expressions by z^3 and you will have z^{-1}, z^{-2} and z^{-3} terms. Then determine all the a and b coefficients as detailed in the example macro. The circuit in LTspice is shown below:

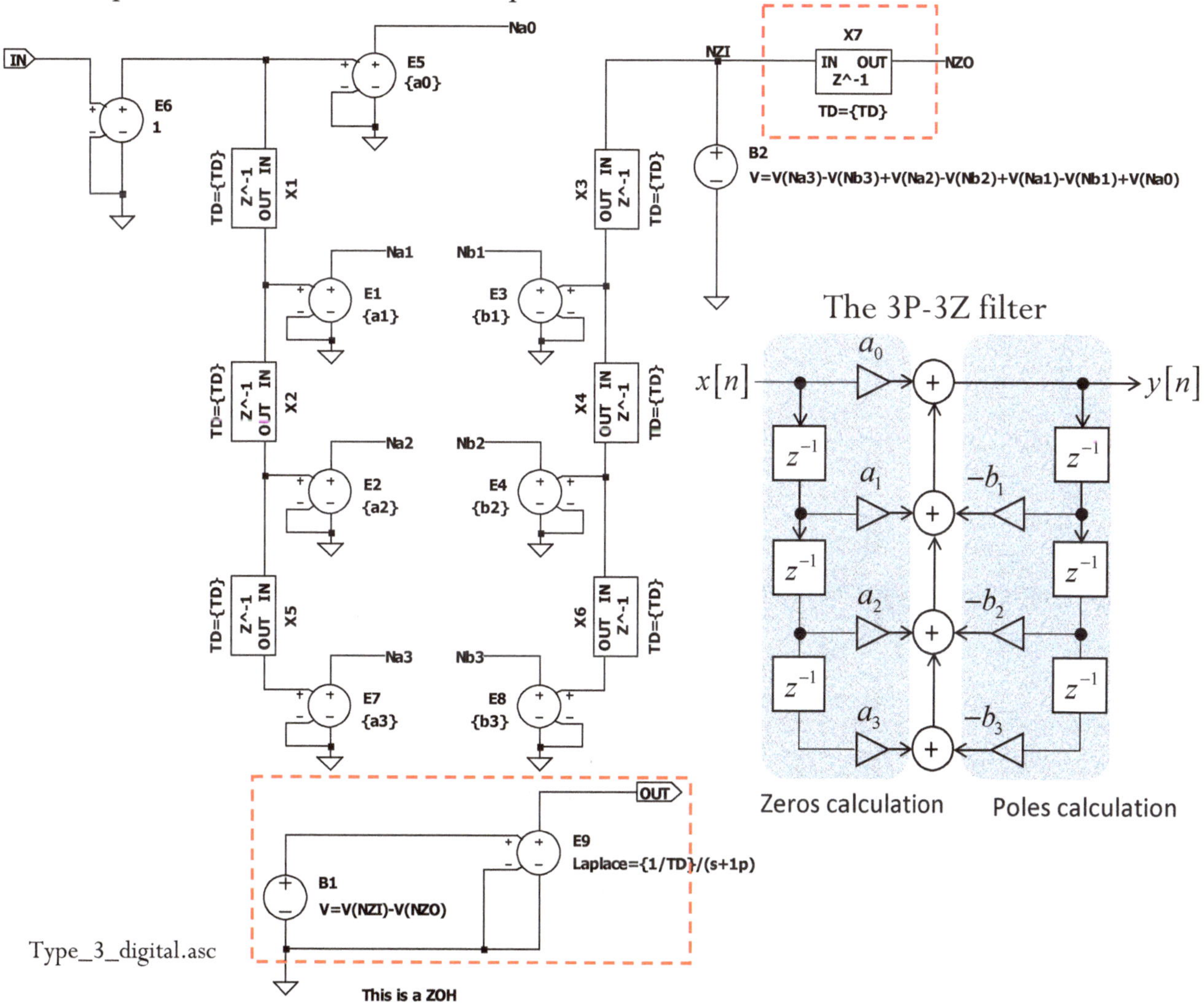

Type_3_digital.asc

The simulation circuit involves an auto-biasing circuit which delivers exactly 12 V and keeps the error voltage around an arbitrary level of 2.5 V.

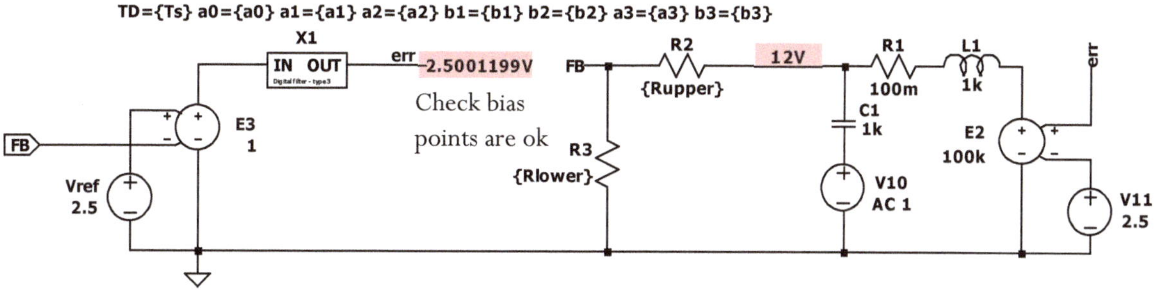

A Digital Type 3 Compensator - II

THE MACRO I wrote to determine the coefficients is quite complex – and too large to fit here – but it is very convenient to change the gain and phase boost on the fly, and see the Bode plot changing accordingly.

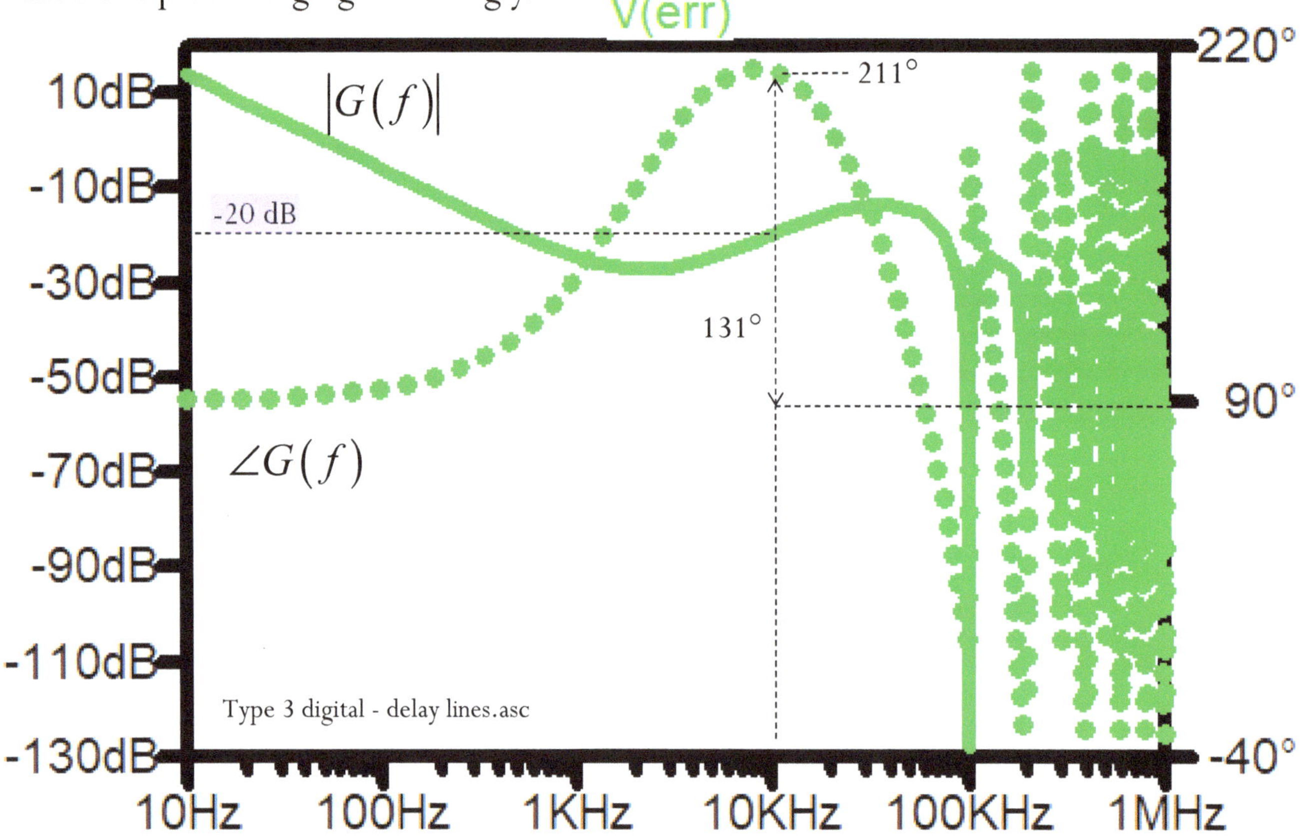

In this example, I used the *k* factor to place two coincident poles and zeros, but you can also involve a manual placement as shown in some of the examples. Here, there is a 20-dB gain excess at 10 kHz and a 130° phase boost is required. Once the simulation is over – and providing you have included a *.options logparams* statement – pressing <CTRL> + L opens the log files and all coefficients are printed.

a0 = 5.627724947261044e-01
a1 = -4.870191878735944e-01
a2 = -5.602232569999946e-01
a3 = 4.895684255997040e-01
b1 = -1.341179593026222e+00
b2 = 3.702804717006065e-01
b3 = -2.910087867438459e-02

The Mathcad simulation confirms the magnitude/phase plots in LTspice with the same 20-dB attenuation at 10 kHz. The ZOH affects the phase a little more (10°) in Mathcad though.

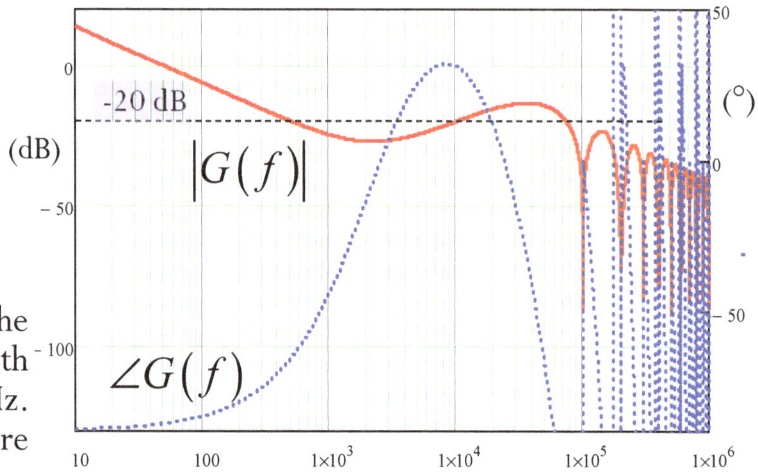

A Digital Filtered PID

THE PID is implemented in digital control and control programs often accept native k_p,k_i and k_d parameters. As usual, before blindly typing the parameters and see their effect on the loop characteristics, it is advisable to independently plot the ac response of the digital implementation. Again, this is quite easy to do with delay lines mimicking z^{-1}.

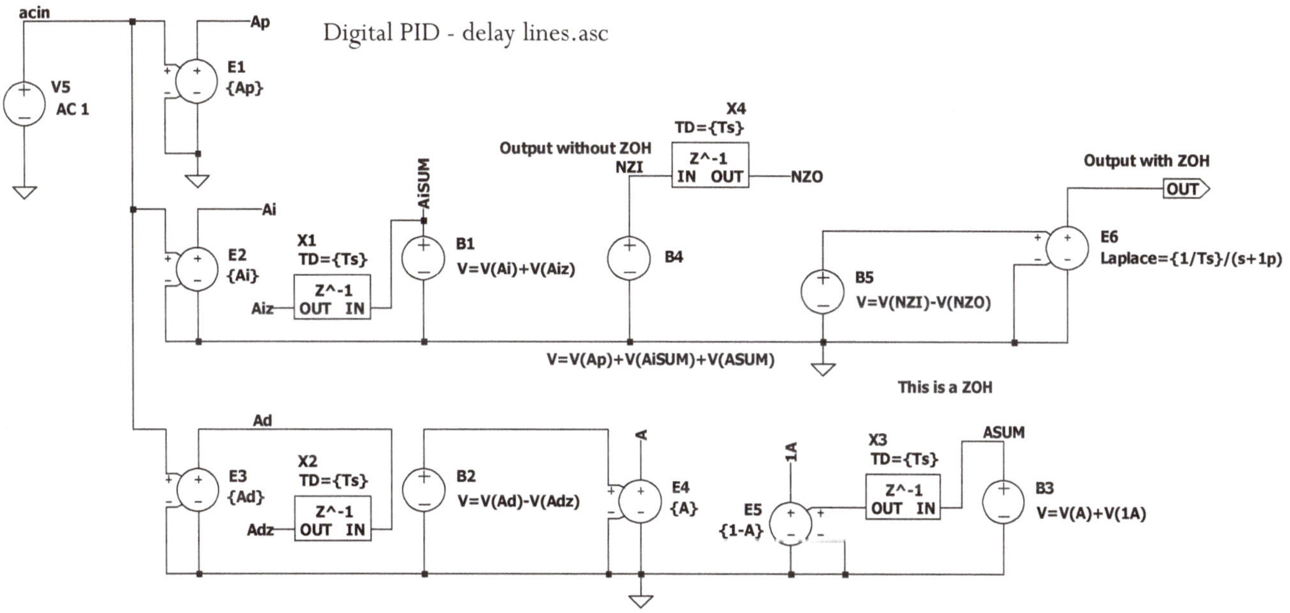

In this macro, I used the raw PID coefficients but you could easily adapt the sheet to enter the poles and zeros you want, then translate them into coefficients (see *PID digital – delay lines – poles zeroes.asc*). This is what I did in the *PID to type 3* example. The plot will compare the output before the ZOH (node NZI), after the ZOH (node OUT) and, finally, I have added a Laplace expression describing the filtered PID, fed with the raw coefficients. The sampling frequency is 200 kHz.

```
.PARAM Fs=100k ; sampling frequency
.PARAM Ts={1/Fs}
*
.PARAM kp=2.643
.PARAM ki=2.509k
.PARAM kd=509.6u
.PARAM fp=21.8k ; filtered PID pole
*
.PARAM Td={kd/kp}
.PARAM Ti={kp/ki}
*
.PARAM N={2*pi*fp*Td}
.PARAM Ad={kp*Td/Ts}
.PARAM a={Ts/(Ts+Td/N)}
.PARAM Ap={Kp}
.PARAM Ai={kp*Ts/Ti}
*
```

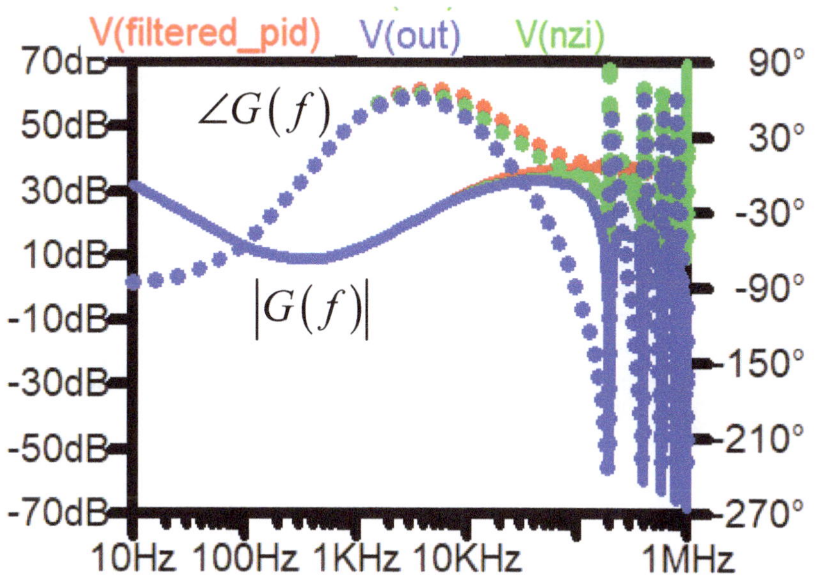

A Digital PI

THE PI FILTER is also implemented in digital control but, this time, you will need to determine k_p and k_i (no k_d). The filter features a pole at the origin for a high dc gain, plus a zero adjusted for phase boost. Add a second pole and you will have a type 2 compensator. Without this pole, this filter is often referred to as a type 2a.

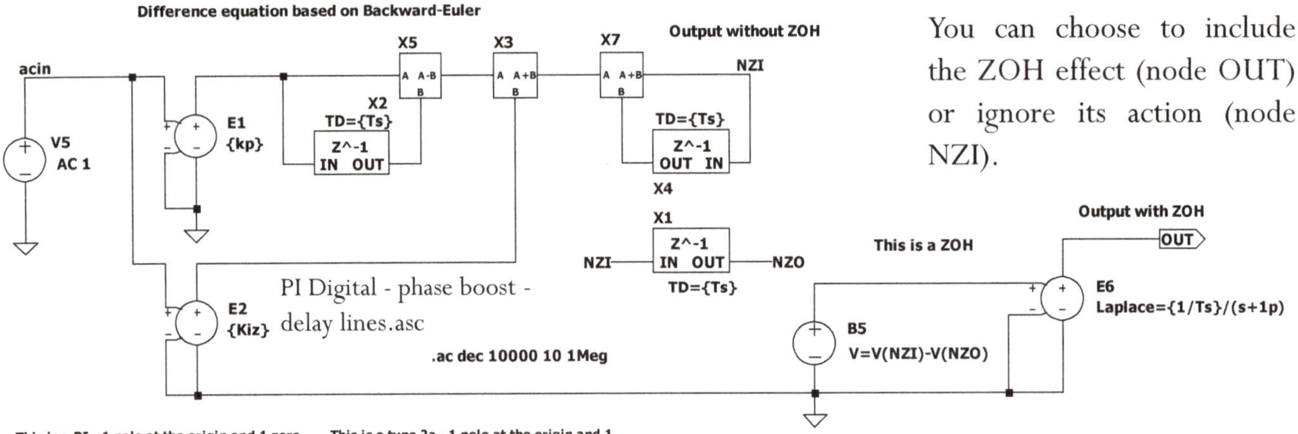

You can choose to include the ZOH effect (node OUT) or ignore its action (node NZI).

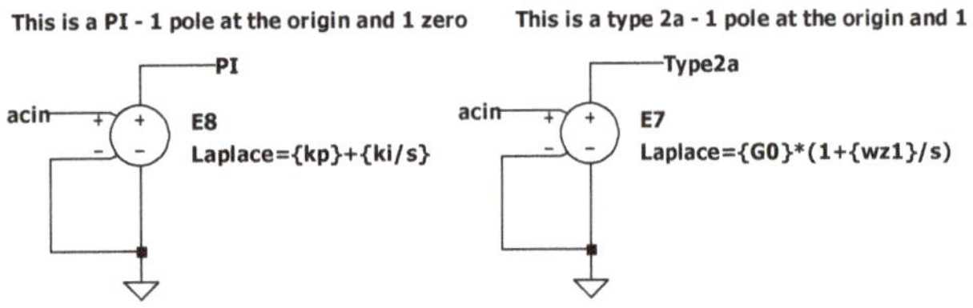

I have added two Laplace expressions for comparing the responses from a continuous-time PI and its type 2a version.

The macro has been adapted to place the zero in relationship with the expected phase boost. The formula differs from the previous implementations because a) there is no phase reversal as with an inverting compensator and b) there is no pole to bring the phase down at high frequencies. However, the sampling effect and ZOH will help bring the magnitude down, naturally helping for gain margin. In this example, the sampling frequency is 250 kHz and the wanted gain is -10 dB.

```
. param Fs=250k
. param Ts={1/Fs}
.param Gfc=10 ; magnitude at crossover
.param PS=-50 ; phase lag at crossover
*
.param fc=1k ; targeted crossover
.param PM=60 ; choose PM at crossover
.param boost={PM-PS-90}
.param fz1={fc/tan((boost)*pi/180)}
.param wz1={2*pi*fz1}
.param G={10**(-Gfc/20)}
.param D={sqrt(1+(fz1/fc)**2)}
.param fp0={G*fz1/D}
.param G0={fp0/fz1}
*
.param ki={fp0*2*pi}
.param kp={ki/(wz1)}
.param kiz={ki*Ts}
```

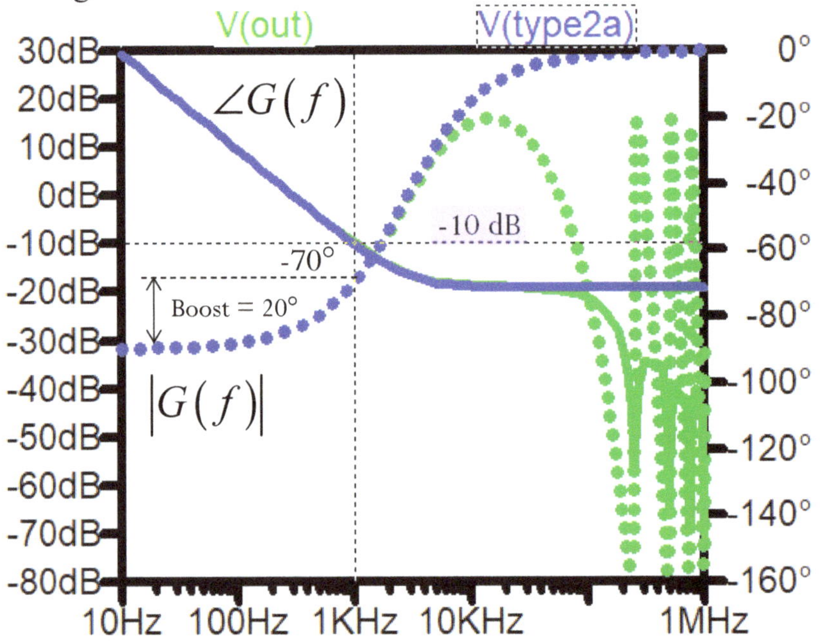

Working with the Examples

NOW THAT WE HAVE detailed most of the subcircuits I have built, we can focus on the real subject, simulating switching converters. The first step is to install the examples and the subcircuits so that LTspice knows where to call them. A purist will object that a subcircuit is a netlist saved in a .LIB file which includes a *.subckt* statement – and I concur with this remark – but for conveniency, I have called a subcircuit, the electrical circuit performing the function I want, saved in a different schematic diagram and associated with a symbol. The result is the same but, with a dedicated schematic diagram, you can have a peek at the internals and modify them easily if necessary.

I have tested all the simulation examples with version 24.1.9. The ZIP file containing the examples, can be downloaded from the *Models to Download* tab on my webpage (https://powersimtof.com/Spice.htm) and extracted following the below structure:

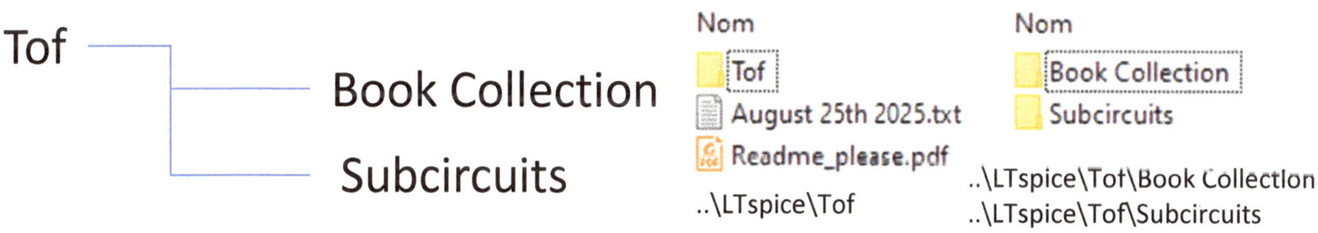

This is how the ZIP file is organized: just extract the Tof main folder to where you save your LTspice circuits and it should do well. Once done, you have to include a new search path for LTspice to find where the subcircuits are located:

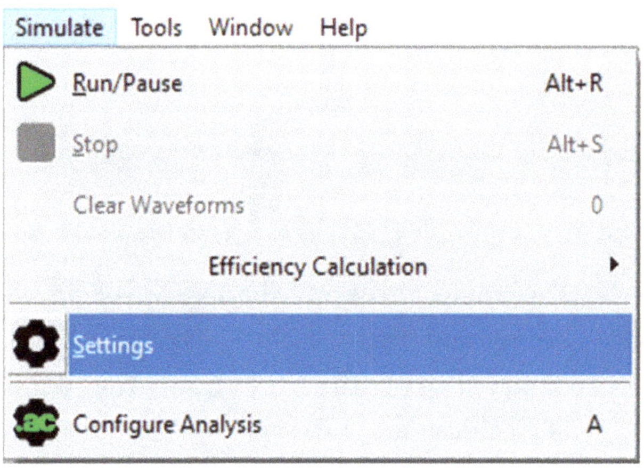

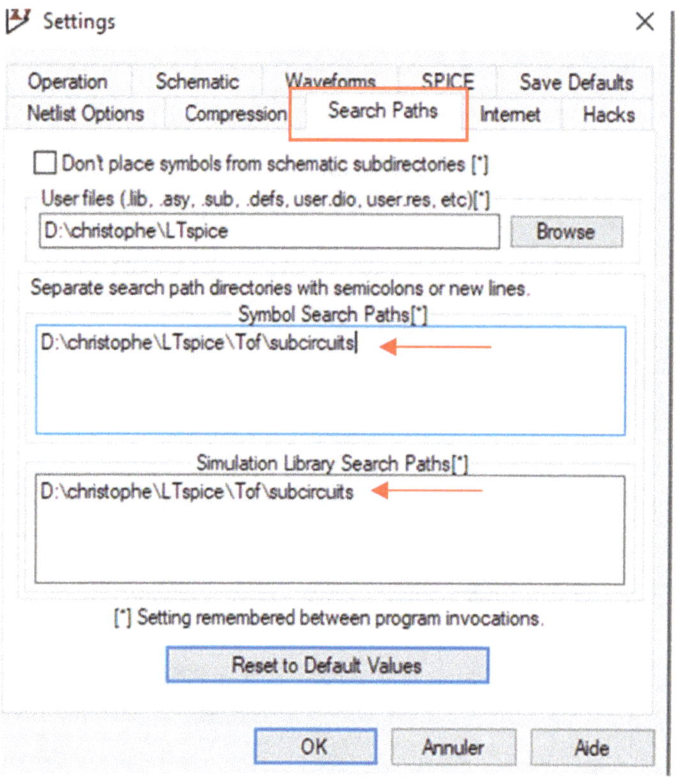

Add this path to your local folder where the LTspice directory has been created. All subcircuits in my examples refer to this location.

Importing Models

IN SOME SPECIFIC cases, you may want to import a model for a component that is not in the library, like a specific MOSFET part or a diode, for instance. LTspice offers many symbols that you can reuse for your component. Let's take the example of a GaN:

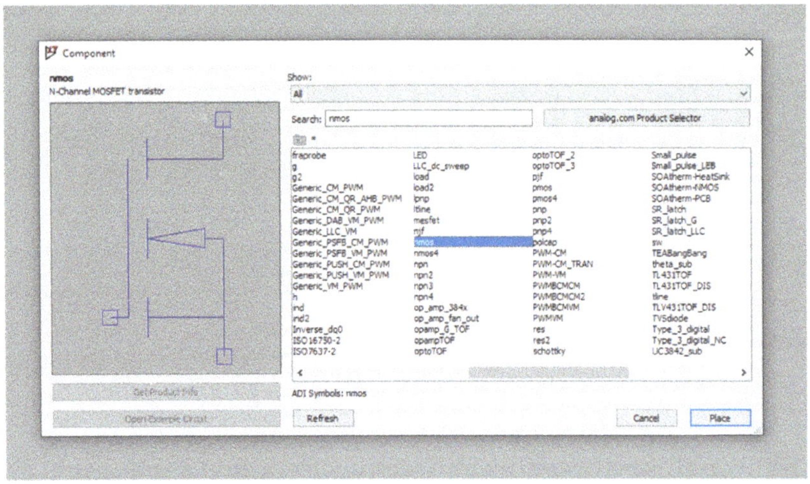

You press P (or push in the ribbon) to invoke the libraries and select *nmos* for a classical 3-pin N-channel device. Place it on the schematic and <CTRL> + right-click on the symbol. Now, you are going to change the prefix from MN to X (a subcircuit). You can also update the *InstName* to Q_1 if you want but that is not mandatory. Then, in the value line, you type in the subcircuit name as it appears in the .subckt of the .LIB file you have downloaded. Once done, you can close the pop-up and add a .LIB statement in the schematic (via the **.t** button) which will indicate the path to where the library file is stored.

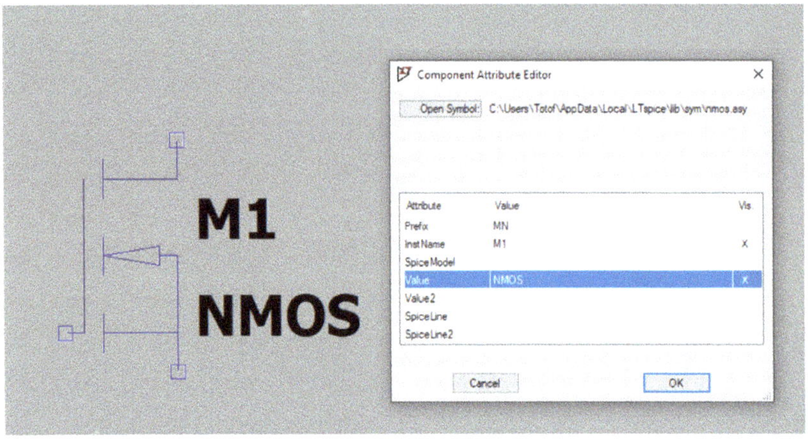

```
.LIB ..\subcircuits\GaN_PSpice_GS-065-011-1-L_L1V4P1.lib
```

```
.subckt GaN_PSpice_GS_065_011 drainin gatein sourcein
```

You could also copy the entire subcircuit netlist and paste it into a control block obtained with the **.t** button. You don't need to use the .LIB statement in this case.

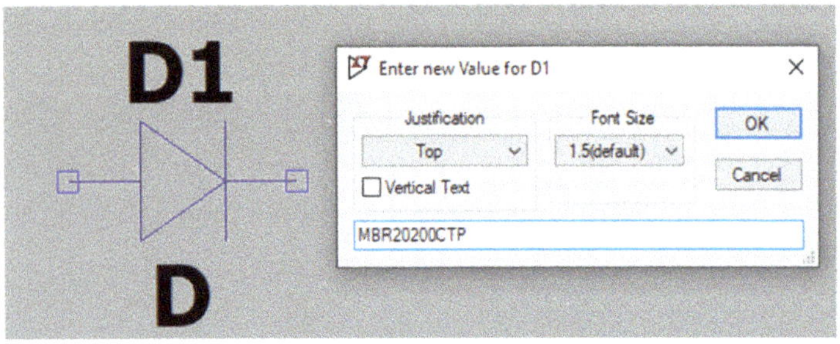

For a simple diode, place it by pressing D and right-click on the D. Type in the model's name and copy its definition into a **.t** block.

```
.MODEL mbr20200ctp d
+IS=7.27826e-05 RS=0.00577989 N=2.51387 EG=0.953093
+XTI=4 BV=200 IBV=0.0001 CJO=4.80195e-10
+VJ=0.2 M=0.422655 FC=0.5 TT=6.49213e-08
+KF=0 AF=1
```

Associating a Subcircuit with a Symbol

SOMETIMES, YOU MAY WANT to import a subcircuit that is described in a book, an article or in a blog. The netlist must be LTspice-compatible and you can save it in a .LIB file for a first approach. Assume we want to include the CCM-DCM subcircuit described in the book *Fundamentals of Power Electronics*. The netlist is as follows:

```
.subckt CCM_DCM 1 2 3 4 CTRL
* D S K A CTRL
*
Gd 0 5 CTRL 0 100u
Rdc 5 0 10k
V1 clpn 0 16m
V2 clpp 0 0.97
D10 clpn 5 DN
D20 5 clpp DN
Vd 1 1x 0
BEt 1x 2 V=(1-V(u))*V(3,4)/V(u)
BGd 4 3 I=(1-V(u))*I(Vd)/V(u)
BEu1 100 0 V=V(5)
BEu2 200 0 V=V(5)*V(5)/(V(5)*V(5)+2*{L}*{FS}*I(Vd)/V(3,4))
D1 200 u DN
D2 100 u DN
RD u 0 100k
.MODEL DN D VFWD=10m
.ENDS
```

Parameters in curly braces

We will have to create a symbol from scratch and link it to the LIB file. Launch LTspice and under the File menu, select *New Symbol*.

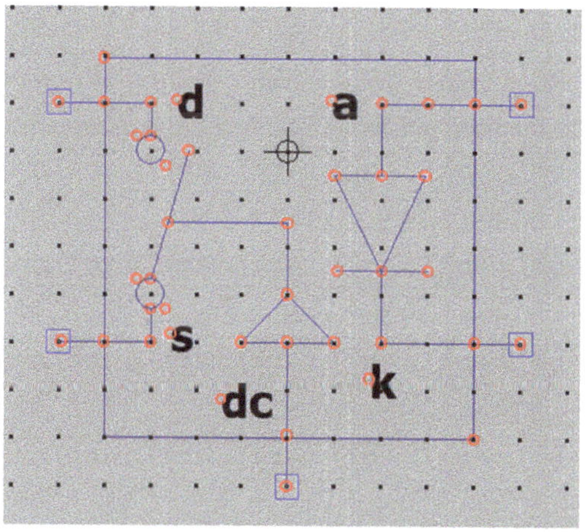

Using the *Draw* sub-menu, build your symbol the way you want by associating a rectangle with lines. Then add pins by pressing P (or Add Pin/Port in the Edit menu). Start with the first pin in the subcircuit, i.e. the drain. It will be pin number 1. I select *Not Visible* for the label and add one with **.t** button, with a smaller font. Continue by adding pins in the order given by the subcircuit (S, K, A and CTRL) and the last pin holds number 5. Once done, save the symbol in the subcircuits directory with a .ASY extension.

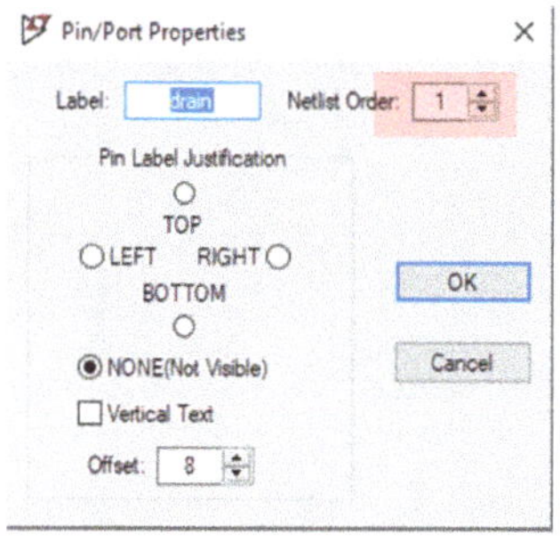

Place the symbol on a sheet, then right-click to feed the attributes as in the figure: X1 for the prefix, U for the name and then pass the subcircuit name in Value. Then add the parameters in Value2: all set!

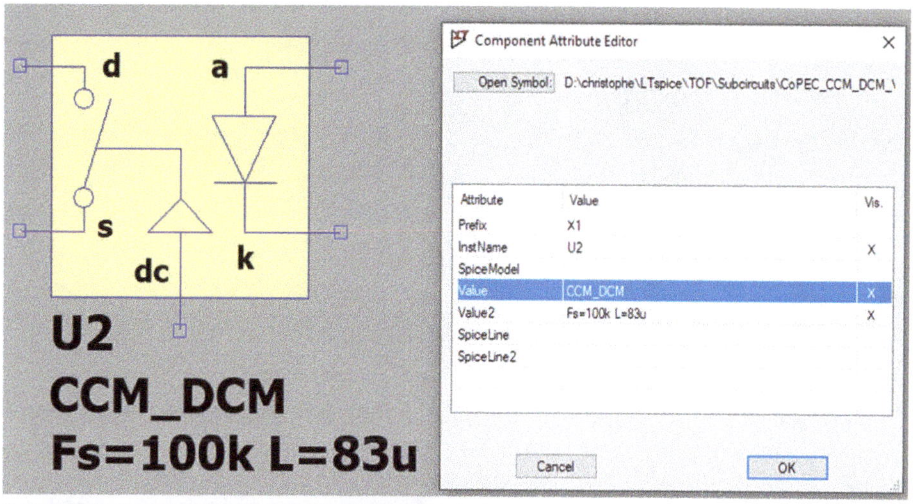

.LIB ..\subcircuits\CoPEC.lib Add the .LIB line on the sheet

The Buck Converter – VM

THE BUCK CONVERTER is part of the basic switching cells, together with the boost and buck-boost converters. It can be operated in voltage- or current-mode control, usually at a fixed switching frequency, but I have seen boundary-mode types as well. There are different ways to simulate a buck converter. The fastest and easiest one, is the *averaged* model. In this approach, there is no switching component and the circuit simulates fast in .AC and .TRAN analyses. As SPICE linearizes nonlinear equations around an operating point – which you need to validate – you can obtain an ac response and characterize a power stage before thinking of an adequate compensation strategy.

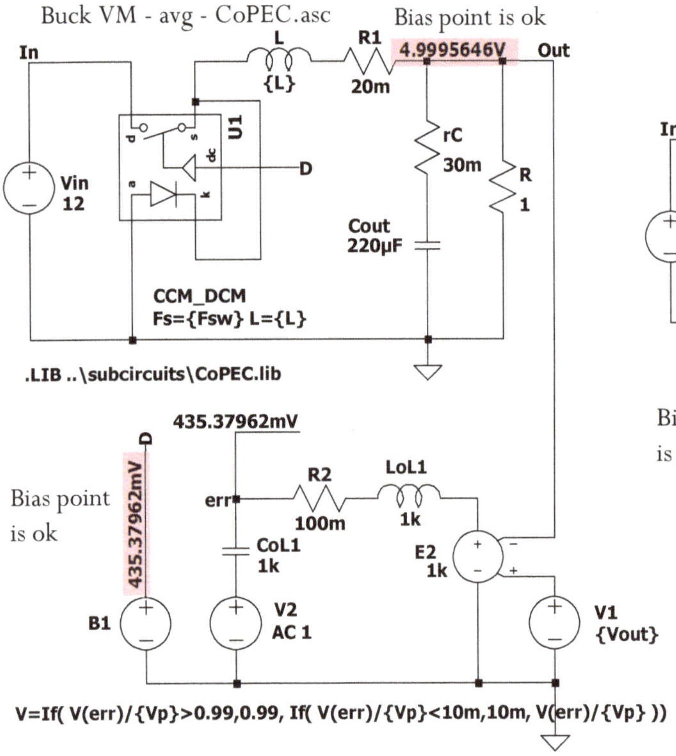

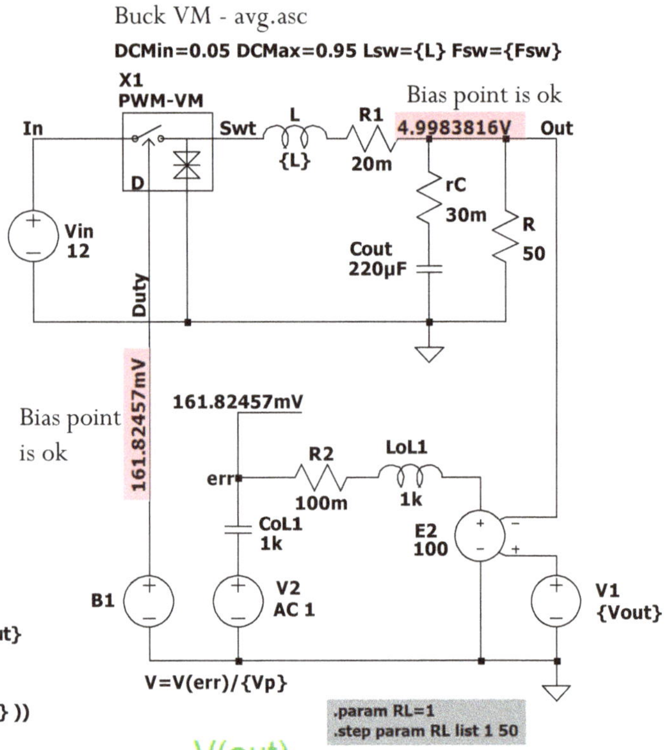

You can use the auto-toggling DCM-CCM model from CoPEC or the one built around the PWM switch model. By stepping the load resistance from 1 Ω (CCM) to 50 Ω (DCM), the auto-bias circuit adjusts the duty ratio (1 V = 100%) and keeps V_{out} at 5 V (the bias points confirm it). Plotting node *Out* gives the control-to-output transfer function for both operating modes.

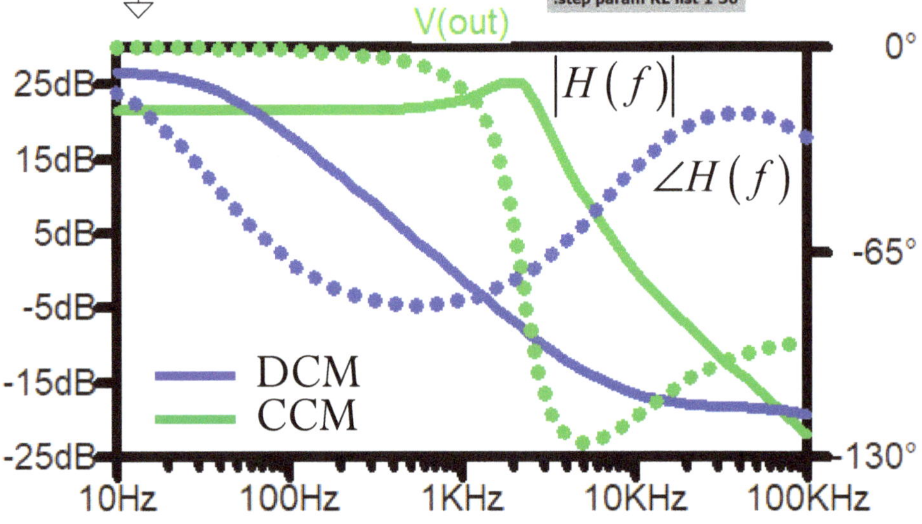

The transition to DCM or CCM is based on the line/load conditions

The Buck Converter – CM

THE BUCK CONVERTER can also be operated in current-mode control where the error voltage controls the inductor peak current and, *indirectly*, the duty ratio. In this mode, a converter operated in CCM becomes a 3rd-order system, with 1 pole in the low-frequency range and two coincident sub-harmonic poles located at $F_{sw}/2$. These poles need to be damped by a compensating ramp, an artificial sawtooth either subtracted from the feedback setpoint or added to the current-sense information. The polarity of the ramp is different between the two methods, but results are identical: you reduce the internal current loop gain and improve stability. Various averaged models exist to simulate a CM power stage but, in my opinion, none surpass the CM PWM switch model developed by Vatché Vorpérian in 1990 [5]. This model works in ac and transient analyses and predicts subharmonic oscillations in CCM. I have introduced an auto-toggling version of this model in 2008 and it has been the object of translations to LTspice by various authors. I use the one ported by my friend Didier Balocco which converges well. There is one version for ac simulations and another one for transient exercises. For the latter, the internal resonant capacitor is not present for convergence reasons.

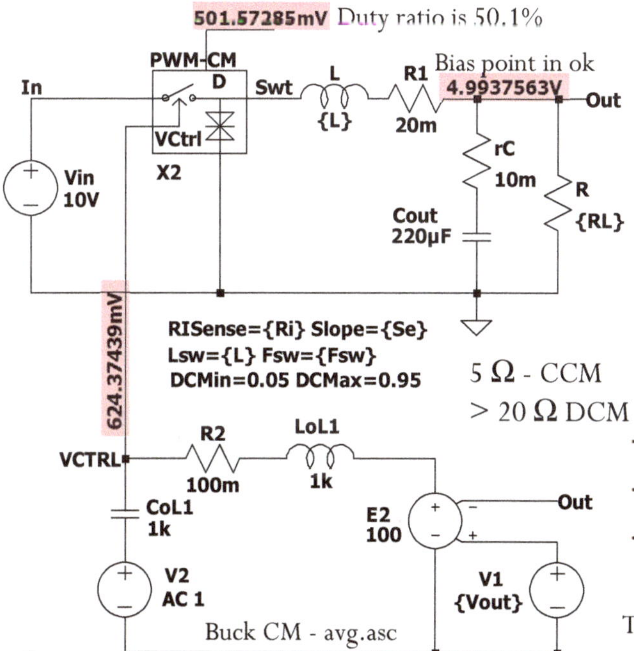

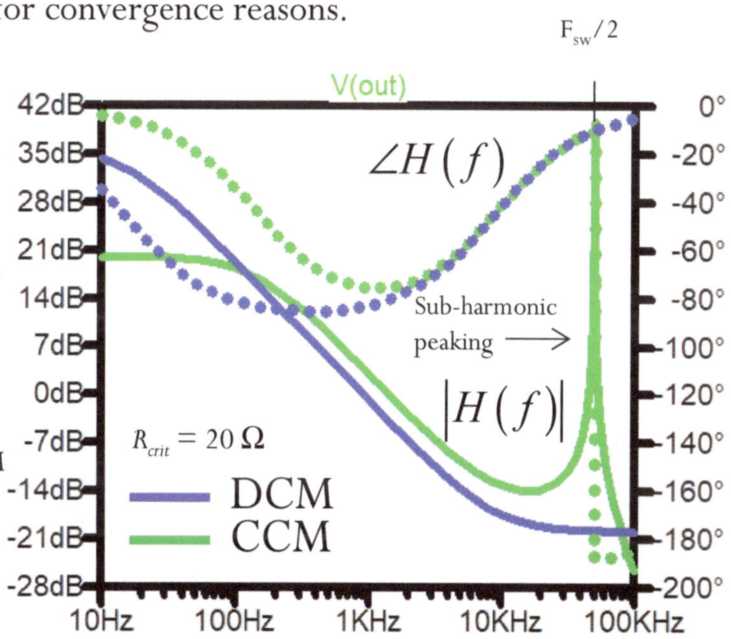

The transition to DCM or CCM is based on the line/load conditions

```
*
.param L=50u ; inductance value
.param Vin=10 ; input voltage
.param Vout=5 ; output voltage
.param Fsw=100k ; switching frequency
.param Ri=0.5 ; sense resistance
.param Sn={(Vout-Vin)*Ri/L} ; on-time inductor current slope
.param Mc=1.0; amount of slope compensation
.param Se={(mc-1)*Sn} ; external ramp level for meeting Mc
*
```

You need to pass a few parameters to the model. R_i is the sense resistance and S_e is the compensation ramp in V/s. m_c can be used to automate its calculation: m_c=1 means no ramp (the simulation), m_c=1.5 means 50%.

The Frequency-Response Analyzer

IN THE PREVIOUS examples, we have illustrated the frequency response of a buck converter using an averaged model. In this model, nonlinear equations describe voltages and currents of the power switch and the diode, *averaged* along a switching cycle. When LTspice launches its ac analysis, it will first determine the dc operating point and linearize all expressions around this point. The system is now linear, and the engine initiates an ac analysis with a 1-V ac stimulus.

Averaged models simulate fast because the switching component has disappeared during the *averaging* of current and voltage waveforms. They are a simplified representation of a switching converter and you do not see the higher-order switching effects brought by parasitics. Besides, they do not work for resonant circuits like LLCs where power is transmitted via the switching component and its harmonics, naturally absent in an averaged model.

You can also build a complete cycle-by-cycle switching circuit, and excite the control input to see how this *stimulus* propagates in the network and produces a *response* at the output node. The complex ratio of the response to the stimulus, is the *transfer function* we want. It works with any kind of circuit, provided the excitation level is adjusted to a) always keep the circuit linear and b) inject with a sufficiently-high level so that you can extract the response drowned in the switching noise.

You can either *physically* open the loop – usually ok to extract the power stage ac response – or measure the compensated *open-loop* gain by keeping the dc loop closed and inserting a perturbation in the return path. In this mode, the loop is *closed* in dc – to fix the operating point via the compensator – but *open* in ac. This last option will give you the open-loop Bode plot and let you extract crossover and phase/gain margins.

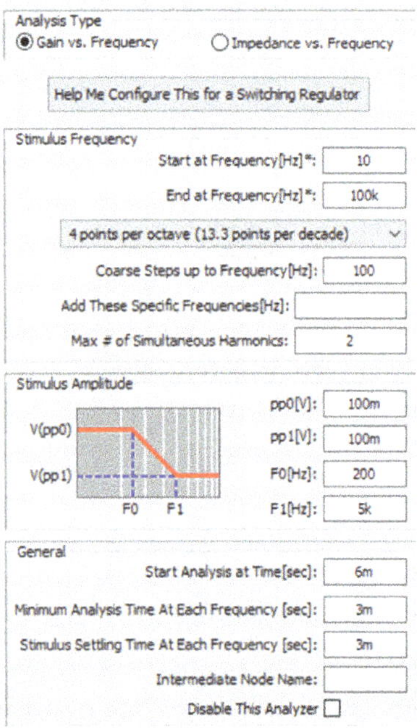

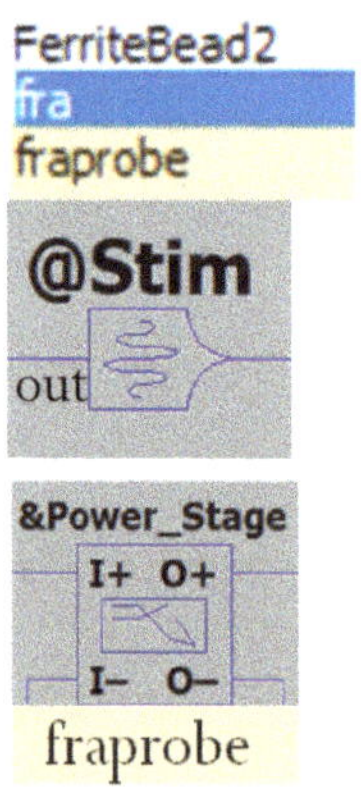

You place the FRA symbol from the component menu. This is the excitation stimulus that contains all the parameters for the ac sweep. You can access the menu by right-clicking on the symbol. I usually sweep between 10 and 100 kHz. For a peaky response, 3 or 4 points per octave give good results, with a total of 2 frequencies applied during the measurement. The modulation amplitude can be fixed for a physically-open circuit but needs adjustment when the FRA is inserted in a closed-loop circuit (to keep away from saturation as the loop gain decreases when approaching the 0-dB crossover).

I have found that analysis and dwell times of 1 ms or more, gave good responses, but smaller values will speed up simulation time. The fraprobe symbol is a differential Bode box used to plot a transfer function between selected nodes. You will see how I have implemented these symbols in the coming examples.

SIMPLIS AC Simulations

LTspice INCLUDES a frequency-response analyzer, and that is a welcome feature. However, as in the laboratory with a FRA, you have to control the amplitude along the frequency sweep to prevent the converter from saturating. If this is done quickly with a real instrument, it can be quite time-consuming with a SPICE engine since you need to go through lengthy iterations. In addition, when it comes down to power factor correction stages, where you start sweeping at 100 mHz or so, it becomes almost impossible to obtain a decent Bode plot of the power stage in a reasonable time. For these reasons, when the LTspice FRA cannot deliver, I recommend to use the free demonstration version of SIMPLIS, called Elements, which lets you directly extract the ac response from a switching circuit in a few seconds. The piece-wise linear (PWL) engine on which SIMPLIS is based, allows to ac-sweep a circuit via a dedicated source, without requiring action from your side to tweak the right amplitude. By determining a steady-state operating point – very precisely, down to the pico volts and amperes – the program keeps the noise floor at an extremely low level and delivers a Bode plot with adequate granularity. It works for fixed-switching-frequency converters, but also for resonant topologies. All the files are available from my webpage.

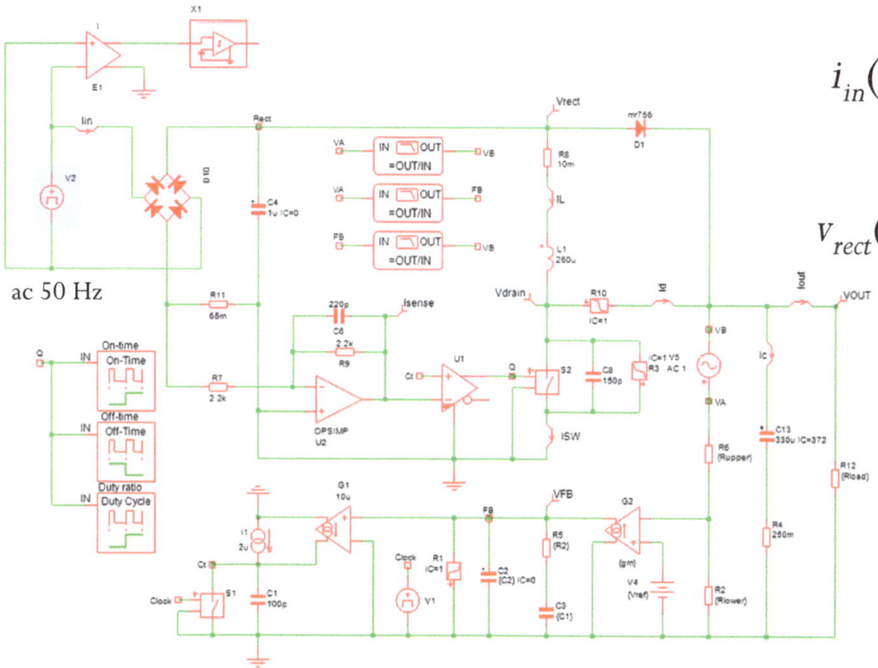

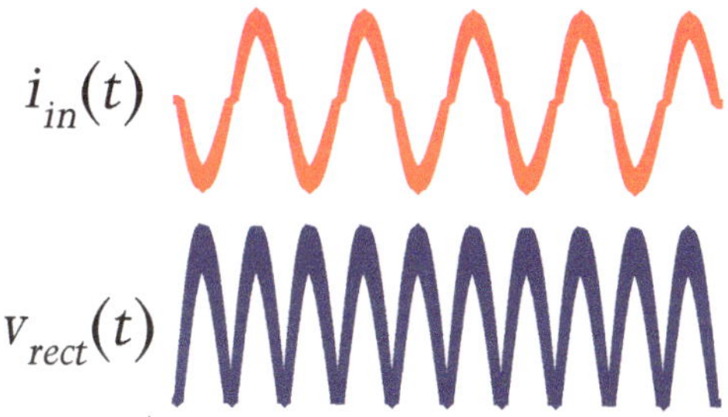

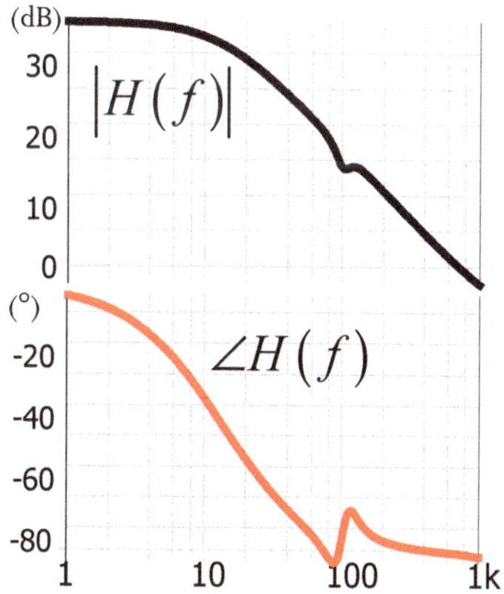

As a unique feature, the control-to-output transfer function of this PFC stage is obtained with the sinusoidal input source still on, as in real operating conditions.

The ac response is obtained after 5 mn on my old machine. It is easy to automate the type 2 calculation with the OTA and quickly see the effects on the Bode plot then the transient response.

A Simple Buck with FRA

THE AVERAGED MODEL is an efficient way to quickly obtain the ac response of a power stage. Another option involves the built-in frequency-response analyzer available with LTspice, which lets you extract the control-to-output transfer function of a *switching* circuit. In this .FRA mode, the simulation can take some time to complete and requires care in selecting the modulation amplitude, especially in closed-loop systems. Here, we will document the simplest possible example with a buck converter driven by a basic pulse-width modulator.

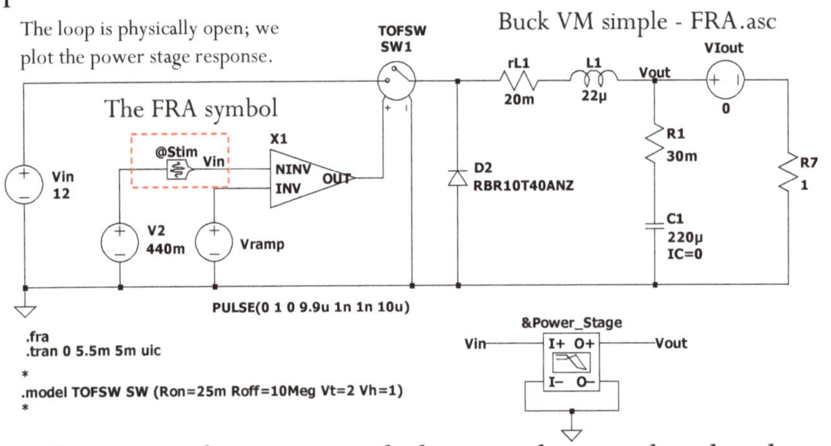

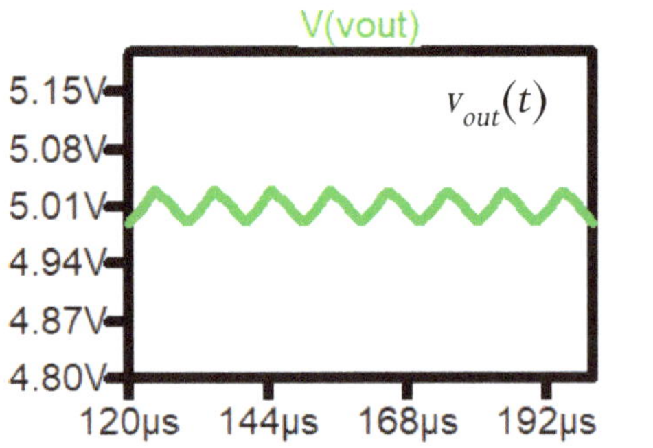

The output voltage is stabilized at 5 V, well before the 6 ms specified in the FRA menu.

In this example – purposely kept at the simplest level – a comparator modulates the duty ratio by comparing the 440-mV dc source with the 1-V sawtooth, leading to a duty ratio of 44%. A transient simulation will confirm the correct output voltage and the time it takes to reach this value. Once this is done, you can run the .FRA analysis:

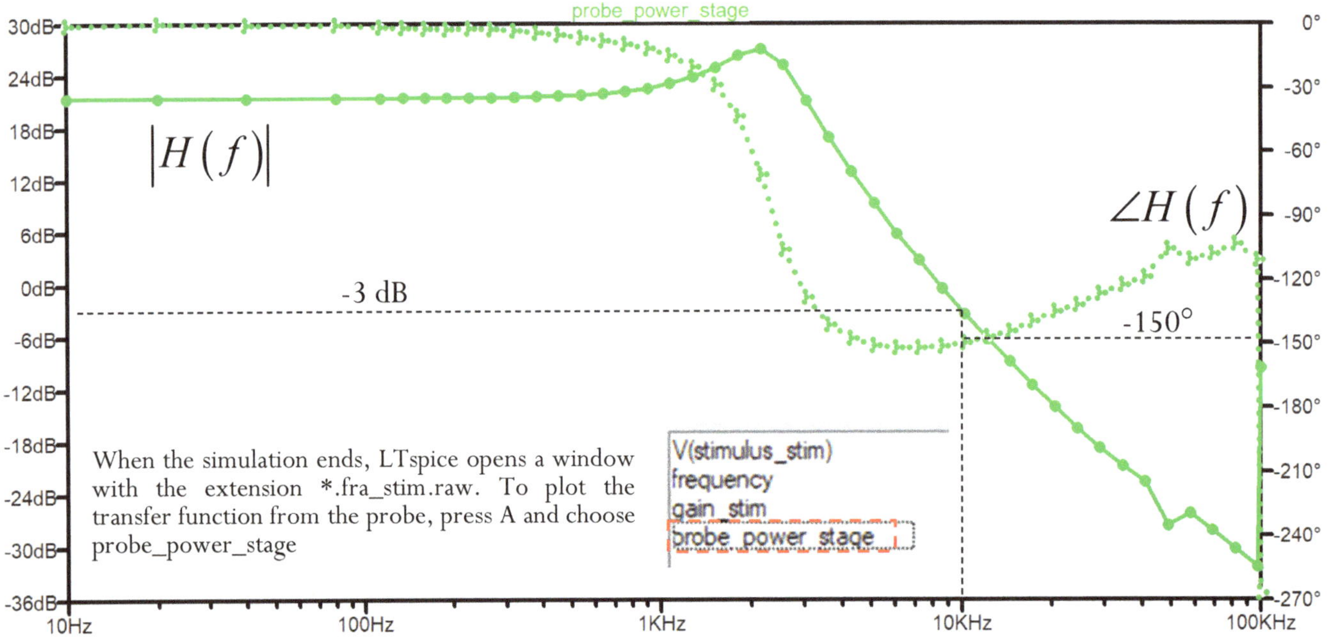

The simulation took 43 s on my old machine to complete, which is acceptable. The granularity is good enough to identify the resonant peak at 2.3 kHz. Read the graph to extract the magnitude and phase at the selected crossover frequency, 10 kHz here.

Compensated Buck with FRA

NOW THAT WE have extracted the power stage response, we can think of a compensation strategy to close the loop. With a CCM buck converter operated in voltage-mode control, we have to cross over, at least, 3-5 times beyond the resonant frequency of the *LC* network, i.e. 10 kHz in this example. We read the Bode plot of the power stage (previous page) and extract magnitude and phase data at 10 kHz: -3 dB and -150° respectively. We will tune our compensator to exhibit a 3-dB gain so that when added to the power stage magnitude, it will give 0 dB at 10 kHz, our wanted crossover. However, the modulator section now includes a divide-by-3 component (-9.5 dB), specifically to increase the dc bias at the op-amp output. The total attenuation is now -12.5 dB and we pass this value to the type 3 macro in the schematic. The phase boost will be determined to 130° for a 70° phase margin.

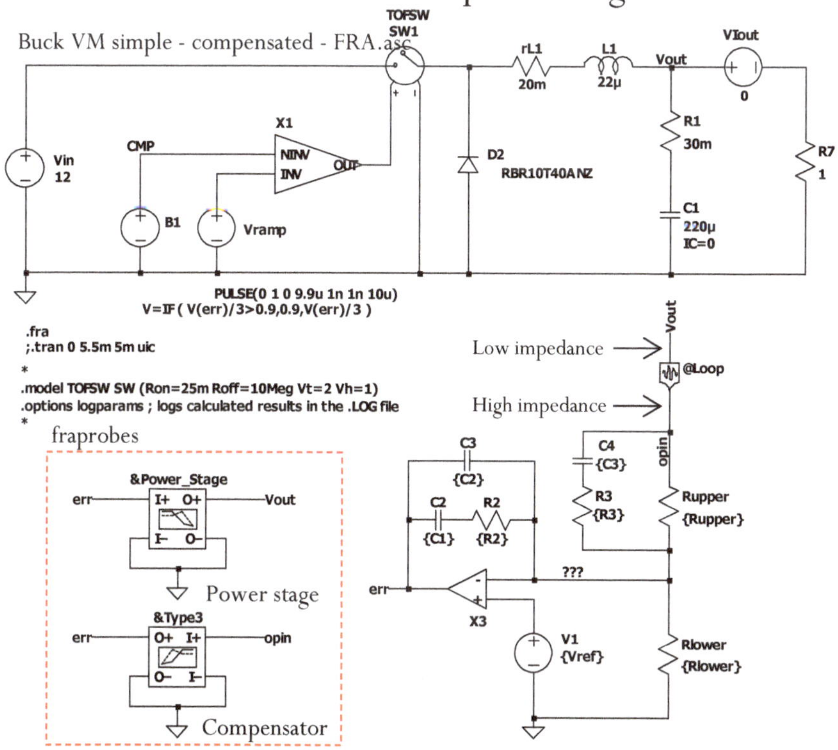

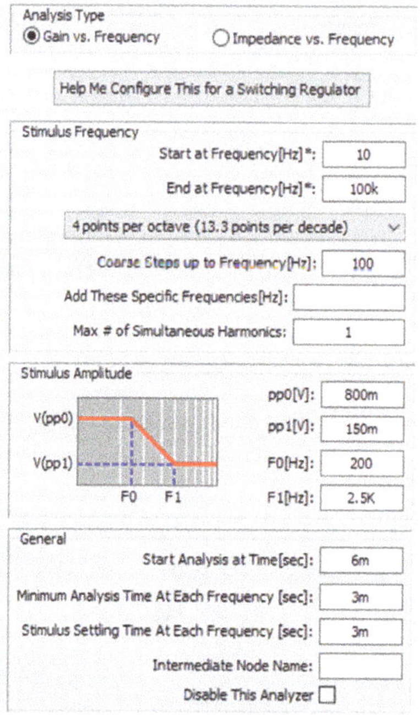

The stimulus amplitude is not fixed but accounts for the decreasing loop gain.

I have placed two zeros at 2 kHz and one pole at half the switching frequency. The second pole is adjusted by the macro to tailor the phase margin to 70°. The stimulus, this time, is inserted in the return path. For correct measurements, the upper point should be connected to a low impedance node (the output V_{out} is perfect) while the injection point should be of higher impedance. The stimulus inserted in the return path is seen as a perturbation by the loop, which will try to eliminate it. To obtain an adequate amplitude on V_{out}, you need to increase the modulation amplitude where the loop gain is high, at low frequencies. However, if you keep the same level, when the loop gain decreases as it approaches f_c, you may saturate the converter: reduce the amplitude but not too much otherwise the plot will be too noisy. This is where you may need a few iterations before you find the right levels, what can be time consuming.

Open-Loop Gain and Transient Response

NOW THAT THE MACRO is fed with the values extracted from the control-to-output Bode plot, we can check the response of the compensated converter and see how it behaves when subjected to a transient step. The *fraprobes* I have installed let me plot the responses of the power stage and the compensator. As you can see, the phase margin is good with 64° – we wanted 70° but the op-amp is imposing its ac response here – and the crossover frequency approaches our 10 kHz target. The simulation time is 104 s.

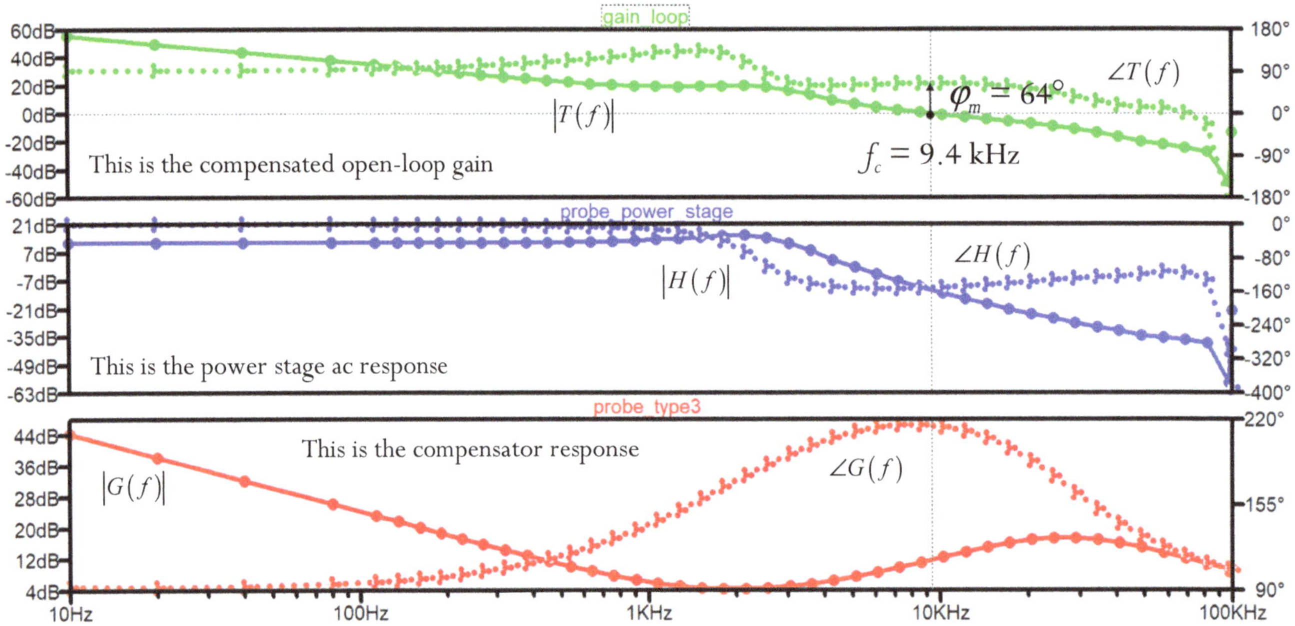

We can test the transient response by stepping the resistive load. {Rload1} and {Rload2} are calculated so that the converter is first loaded at P_{out}·(1-step) then {Rload2} comes in parallel with {Rload1} to obtain 100% P_{out}. The response on V_{out} is excellent.

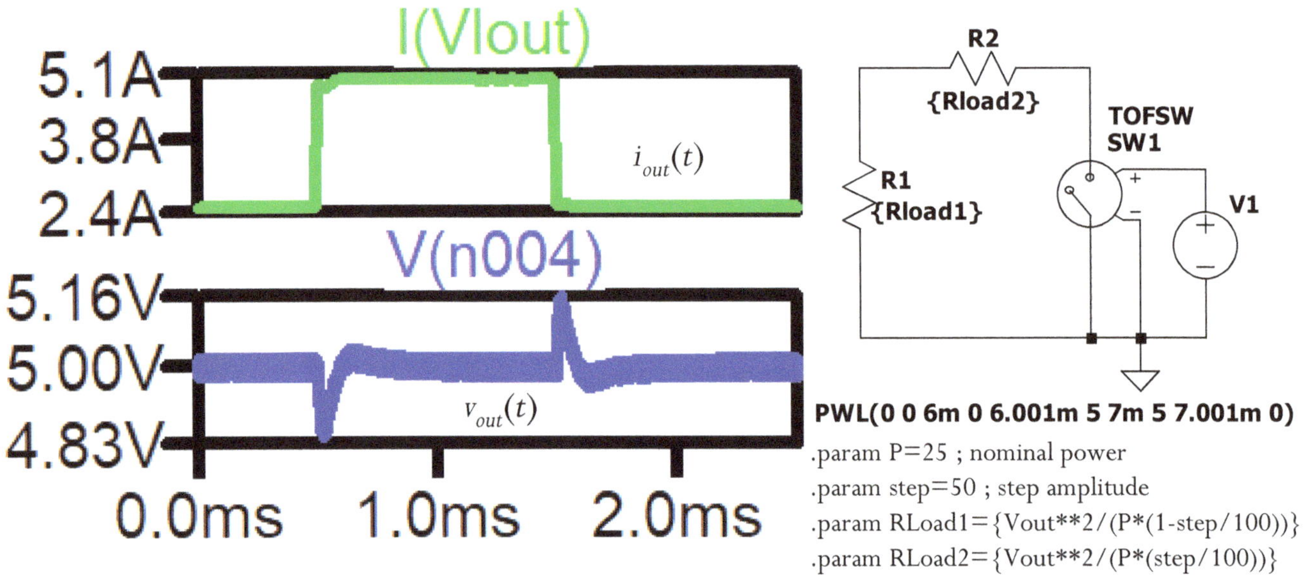

Buck VM simple - compensated - step load.asc

A Cycle-by-Cycle VM Buck

NOW THAT WE HAVE seen how to stabilize a simple buck converter, we can look at a more ambitious synchronous-rectified project featuring our generic controller. I used a second switch SW2 in parallel with diode D_1 to perform the synchronous work. If you want a simple single-switch buck, just disconnect DRVB and ground SW2 control pin. I have used a generic diode and simple switches, but you can of course insert real models after you have confirmed this simple version meets your goals.

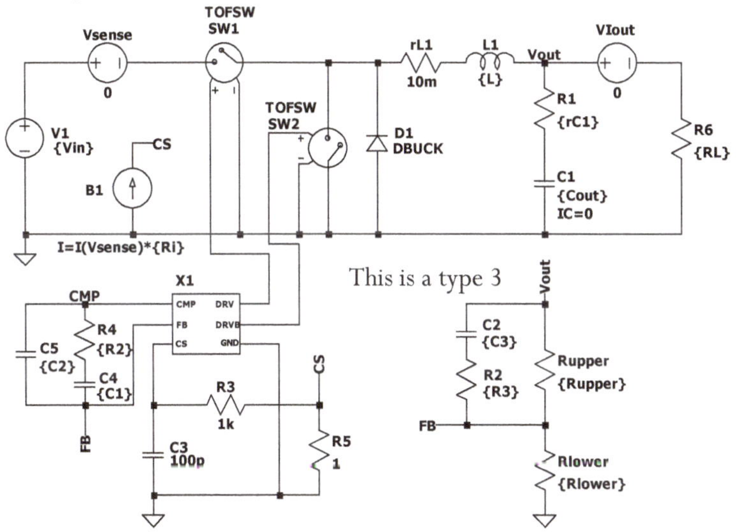

Buck Synchro VM - analogue - cbc - automated - step load.asc

```
.param H0={(RL/(RL+rL1))*Vin/Vp}
.param W0={(1/sqrt(L*Cout))*sqrt((rL1+RL)/(rC1+RL))}
.param Q={L*Cout*W0*(rC1+RL)/(L+Cout*(rL1*rC1+rL1*RL+rC1*RL))}
.param Wzb={1/(rC1*Cout)}
.param Wca={fc*2*pi}
.param Wc=1*Wca ; change crossover frequency
.param Magfc={H0*sqrt(1+(Wc/Wzb)**2)/sqrt((1-(Wc/W0)**2)**2+(Wc/(Q*W0))**2)}
.param Argfc={(atan(Wc/Wzb)-atan2(Wc/(Q*W0),1-(Wc/W0)**2))*180/pi}
```

The 0-V dummy source V_{sense}, senses the switch current and references it to ground with B_1. The output of this source is filtered by R_3/C_3 and goes to the current-sense input for cycle-by-cycle current protection. This function is silent in normal operation, but nicely clamps the inductor current at start-up.

In this example, I have added lines in which I compute the small-signal parameters of the buck stage operated in VM.

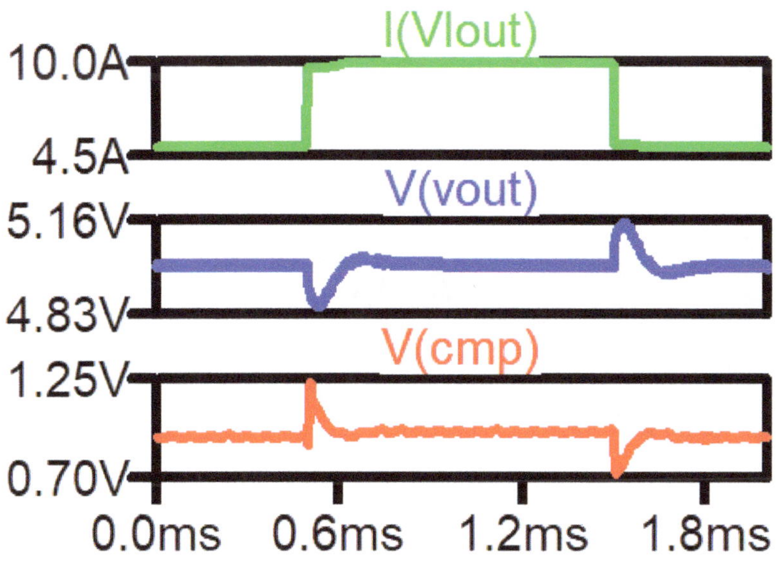

The transient response is very stable and fast ($f_c = 6$ kHz)

That way, I do not need to resort to an averaged model to obtain the control-to-output ac response and I can easily explore different values. Crossover frequency is selected at 3 times the resonant frequency by default, but you can alter it easily and see the effect immediately. Magnitude and phase determined at that frequency, are passed to the type 3 compensator macro.

A Cycle-by-Cycle CM Buck

NOW OPERATED IN CURRENT-MODE, the schematic diagram does not differ much from the previous example. This time, the inductor information is active cycle-by-cycle and feeds the CS pin via a small R_3C_2 filter. In this network, the series resistance labeled R_3 sets the amount of slope compensation: when of low value, there is no compensation at all. When this resistance increases, you mix the current-sense voltage information with the internal ramp and it decreases the inner current loop gain, stabilizing the converter.

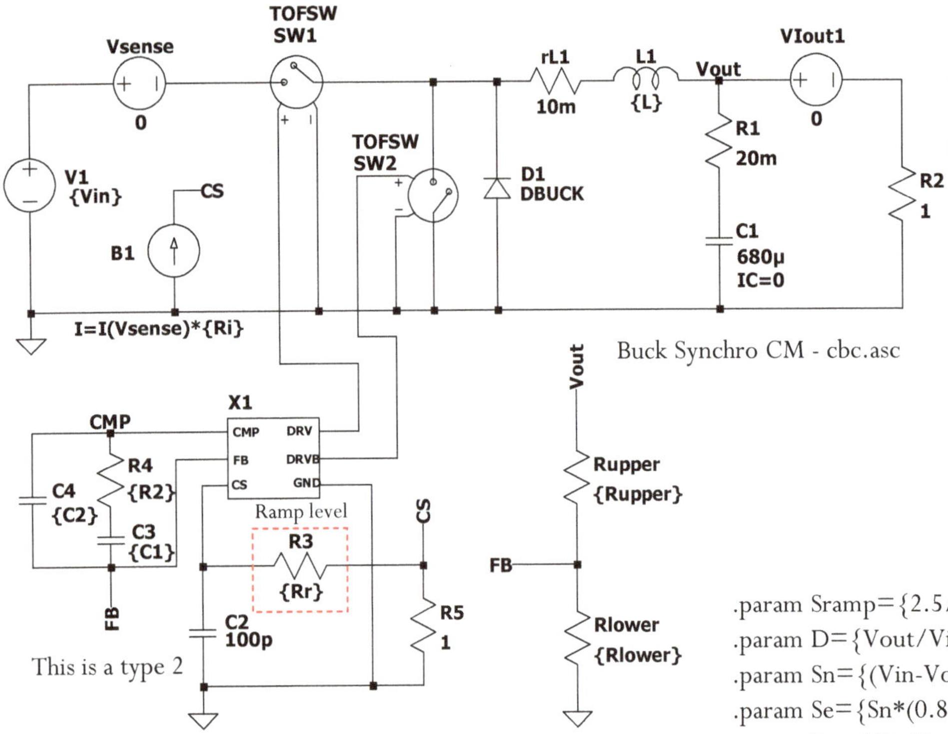

The level of compensation ramp S_e (in V/s) is determined by the quality factor of the resonant poles which is set to unity. The series resistance R_3 is then computed – this is $\{R_r\}$ – accordingly and sets the amount of ramp mixed with the current sense information. In case no ramp is necessary, this resistance is set to 100 Ω and ensures the filtering of the current sense voltage.

```
.param Sramp={2.5/Tsw} ; 2.5 V over Ts
.param D={Vout/Vin} ; duty ratio calculation in CCM
.param Sn={(Vin-Vout)*Ri/L} ; current slope
.param Se={Sn*(0.818/(1-D)-1)} ; external ramp
.param Rr={(Se/Sramp)*19k+100} ; series resistance
```

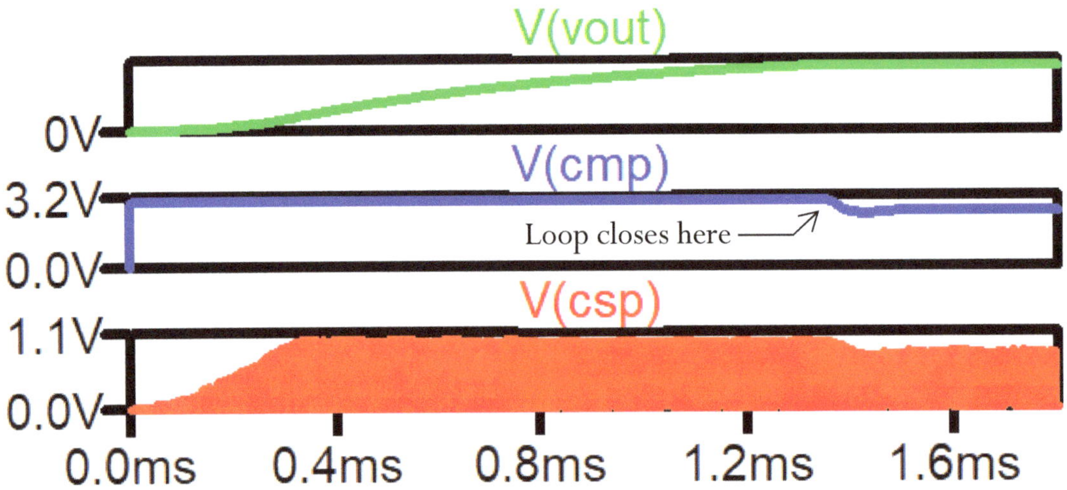

The compensator is a type 2 and goes to the maximum setpoint after the soft-start sequence has elapsed.

The compensation macro uses the *k* factor which works well for a current-mode converter. In this synchronous buck, the converter remains in CCM at all output currents. Its dc transfer characteristic – $V_{out} = DV_{in}$ – is invariant, as well as its ac response.

A Constant On-Time Buck

THE BUCK OPERATED in constant on-time – abbreviated COT – offers interesting characteristics, such as an excellent response to a load step. A COT circuit is built around a single-pulse generator which calibrates the on-time in relationship with the transmitted power. Unlike classical CM operation, the error amplifier sets the inductor *valley* current and not the peak this time. Before looking at the cycle-by-cycle model, it is important to look at the small-signal response and close the loop.

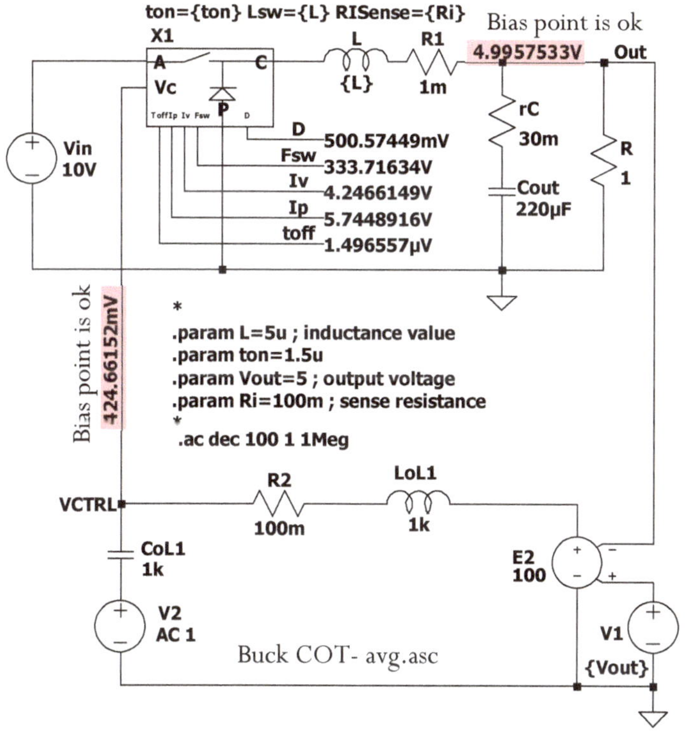

This averaged model determines various parameters once the bias point analysis has completed. The *D* node is the duty ratio (50% in this example), F_{sw} shows a 333.7-kHz switching frequency, the valley and peak currents, respectively noted I_v and I_p, give a ripple of 1.5 A peak-to-peak. Finally, the off-time establishes around 1.5 µs. Together with the 1.5-µs fixed on-time (passed as a parameter to the model), it confirms the switching frequency. As all bias points are within expected values, we can run a frequency sweep.

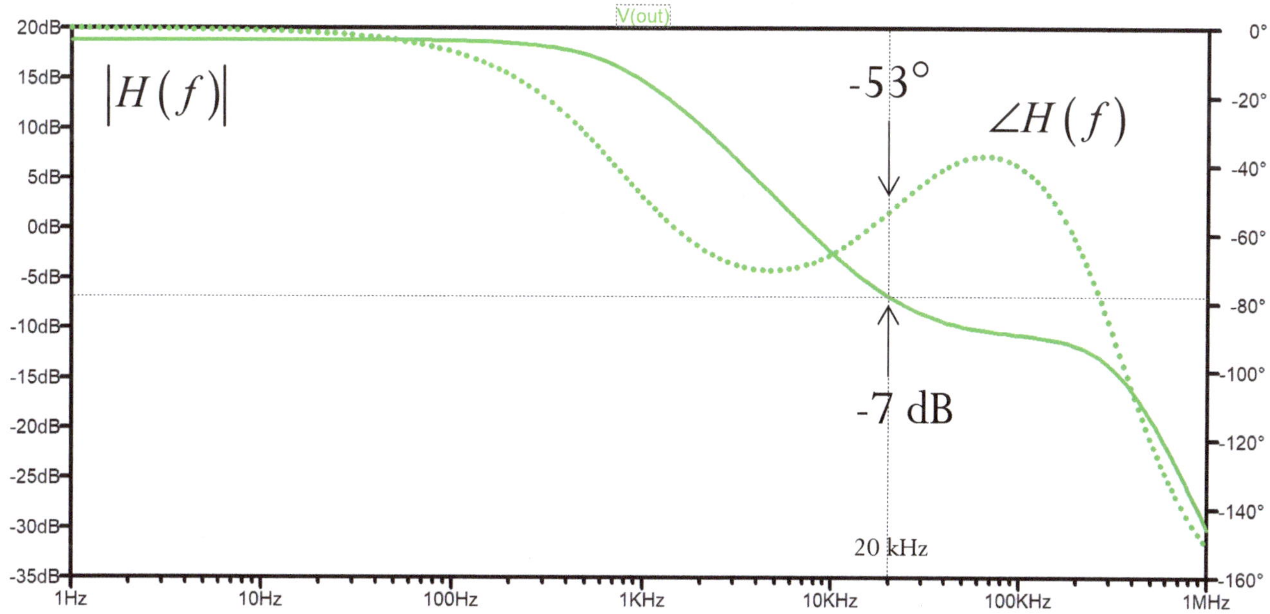

We will push the crossover frequency to 20 kHz in this example. The phase lag is not that high, owing to the capacitor ESR introducing a zero in the considered crossover region.

Compensated COT Buck

NOW THAT WE HAVE the frequency response of the COT buck converter, we can close the loop. The control voltage (pin V_c on the model), is the error voltage output divided by 3 (it brings a 9.5-dB attenuation). This divider increases the dynamics on the op-amp which, otherwise, would operate at too low an output level and its performance could be affected. The compensator will account for this extra attenuation in the macro. As with any converter, the maximum peak current is clamped by {VCS}, the maximum voltage across the sense resistance.

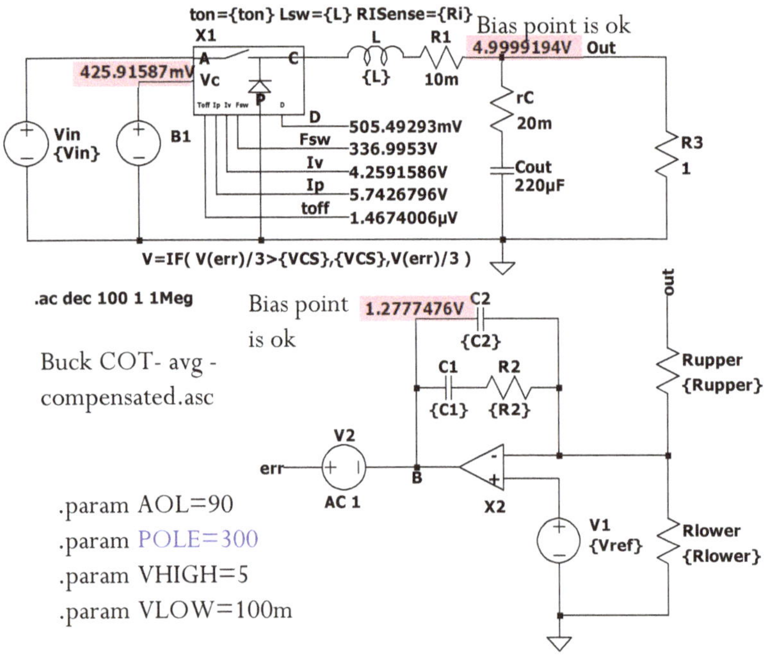

The compensation is classically done with a type 2 circuit. The op-amp specifications have been upgraded for reflecting a higher GBW – the low-frequency pole is now 300 Hz instead of 30 Hz by default – considering a 20-kHz crossover frequency as a goal. When pushing crossover, you always need to check that the op-amp does not impose its own ac response otherwise you won't obtain the wanted compensator characteristics in gain and phase.

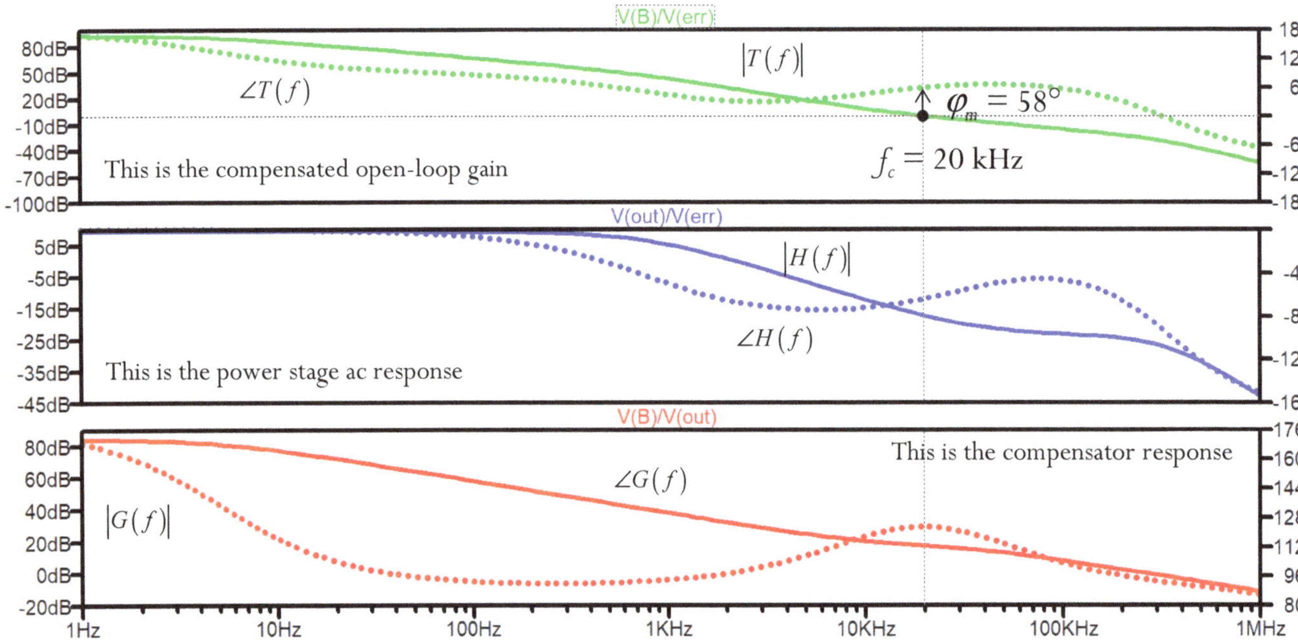

The compensated loop gain shows a comfortable phase margin at 20 kHz and we can expect a good transient response from this compensated COT converter.

Averaged Transient Response

OWING TO THE NONLINEAR equations used in the COT averaged model, we can run a transient test and check if the response meets our requirements. In this 25-W buck converter, I will change the output current from 2 A to 5 A, with a fast slope. What is interesting with this model, is that you can monitor all the interesting parameters, such as the frequency or the duty ratio evolutions during the transient analysis. As usual, you can change the control strategy or other parameters – sense resistance, inductance, on-time – and see the impact immediately, as there is no switching component.

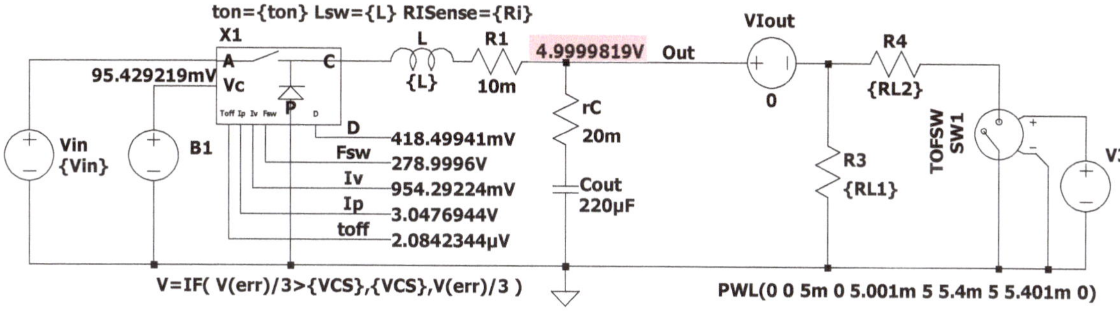


```
.tran 0 5.8m 4.8m
*
.param P=25
.param step=60 ; if step=30, then load is first 70% then goes 100%
.param RL1={Vout**2/(P*(1-step/100))} ; first load
.param RL2={Vout**2/(P*(step/100))} ; final load
.model TOFSW SW (Ron=25m Roff=10Meg Vt=2 Vh=1)
*
```

Buck COT- avg – load step.asc

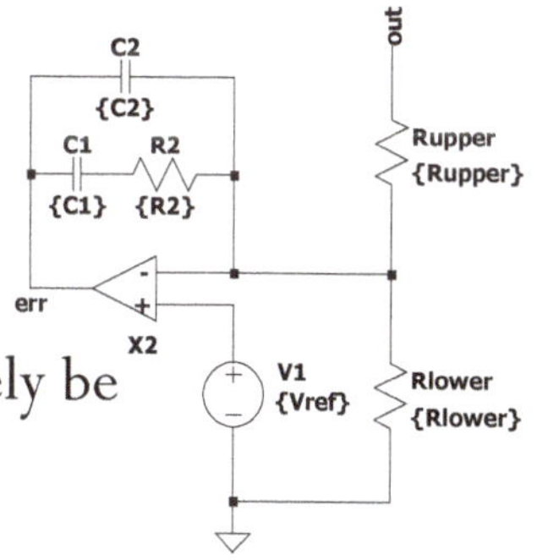

In a practical application, the output capacitor would likely be a MLCC type, featuring a lower capacitance and ESR.

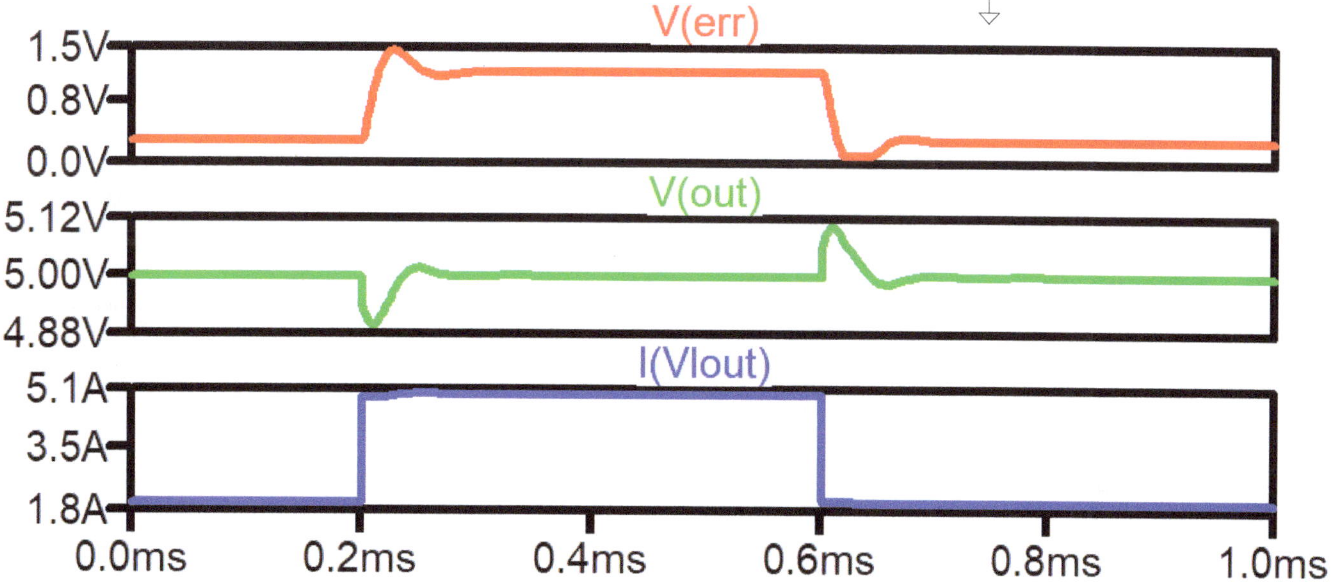

The transient response is very good and shows a quick recovery during and after the load-step event. These COT converters naturally enter standby mode, meaning that the switching frequency decreases in light-load conditions. This is beneficial to the overall efficiency but could also be troublesome when the audible range is entered.

Cycle-by-Cycle COT Buck

FOR THIS EXPERIMENT, I have assembled the complete control section without resorting to a generic model. That way, you can exactly see how the circuit operates. You have the on-time generator in the lower left corner of the schematic diagram. The capacitor is calculated based on the given t_{on} value, 1.5 µs here. A dead time is added to control the synchronous switches, implying a continuous conduction mode operation, even in light-load conditions.

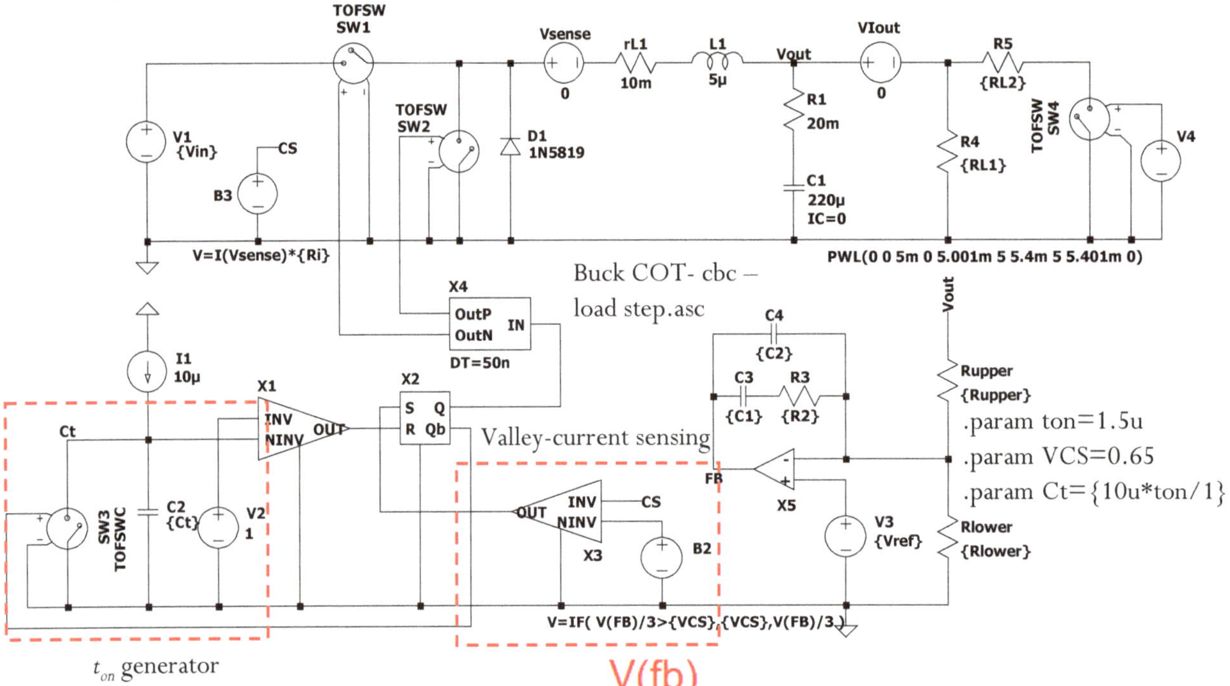

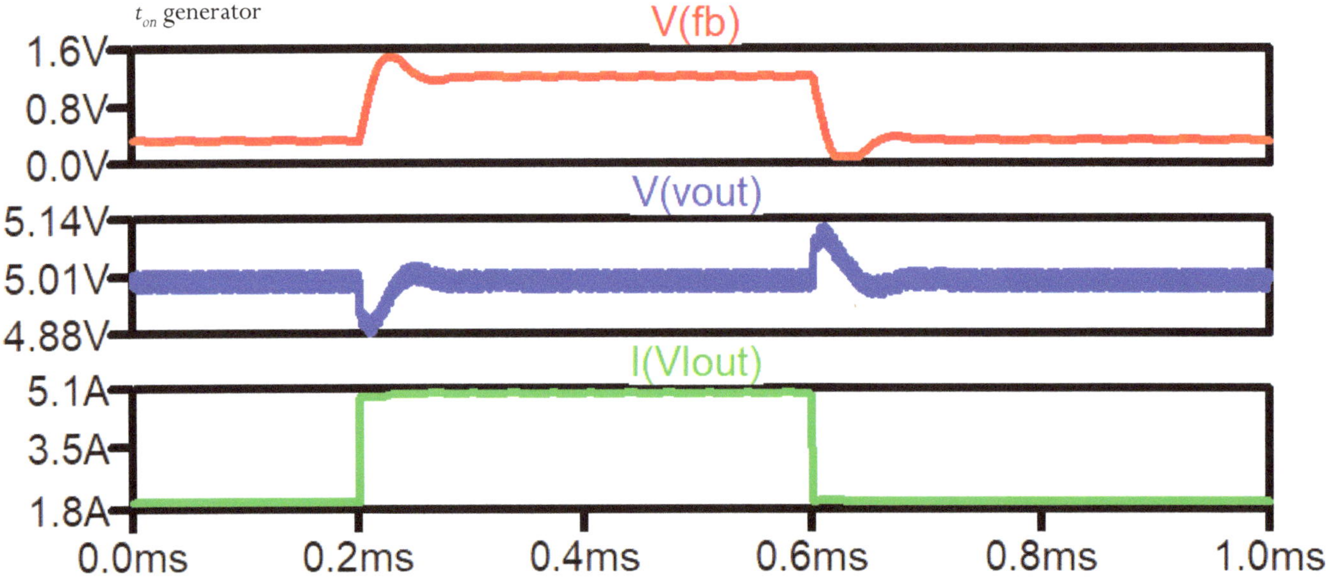

If you compare these curves with those on the previous page, you can see how the averaged model perfectly reproduces the transient behavior of the COT buck converter. Having a validated averaged model, helps run many experiments with it – like changing poles-zeros positions, adopting a smaller inductance – without having the penalty of a lengthy cycle-by-cycle simulation.

A Buck Converter with UC384x

THIS IS AN INTERESTING circuit for experiments only, as the UC384x has not really been designed for driving a high-side switch. Nevertheless, I have built this circuit for the purpose of an APEC seminar during which I was demonstrating loop measurements.

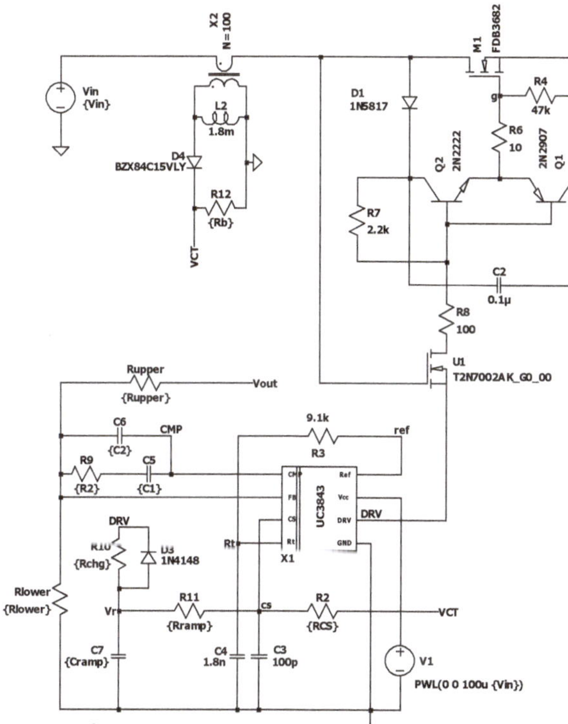

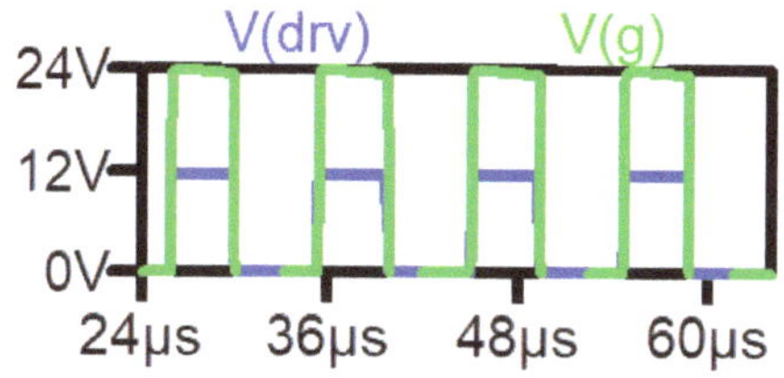

Being located in the high-side, transistor M_1 cannot be driven directly from the ground, unless the DRV pin swings above the source by 10 V or so. One solution is to resort to a bootstrapped circuit where capacitor C_2 is charged at V_{in} via D_1, when D_2 freewheels. Once charged, it can independently supply Q_1-Q_2 to efficiently drive MOSFET M_1. U_1 plays the role of a level translator, controlled by the ground-referenced UC3843. The waveforms confirm a 12-V drive level in green (12 V input, 24 V ground-referenced gate voltage).

Buck CM - UC384x - cbc.asc

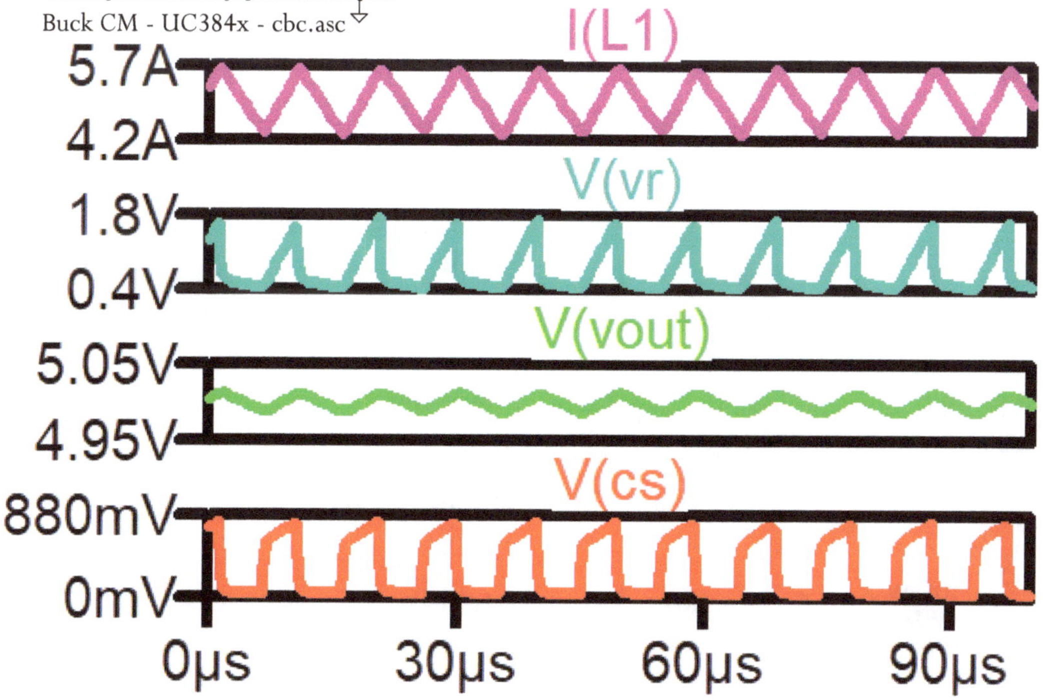

The compensation ramp $v_r(t)$ is not generated from the oscillator sawtooth but from the low-impedance DRV pin. It adds to the circuit robustness by leaving the oscillator untouched. The series resistance {RCS} is computed to set S_e at the desired value.

Owing to current-mode control, a simple type 2 compensator is perfect to stabilize the loop, with the *k* factor method for determining component values. For a type 3, the *k* factor places two coincident pole-zero pairs, which can sometimes lead to conditional stability. In that case, I prefer to manually place poles and zeros, as shown in my macros.

An Averaged CM Forward Converter

THE FORWARD CONVERTER is a buck-derived topology, bringing isolation and scaling via a transformer. Before looking at a cycle-by-cycle implementation, we can first stabilize our circuit by resorting to the current-mode averaged model.

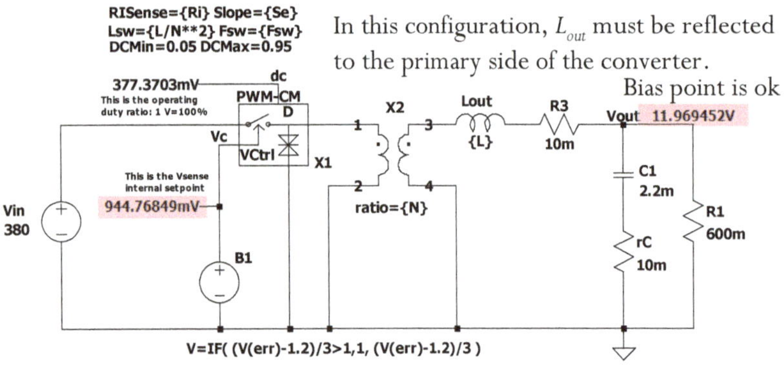

In this configuration, L_{out} must be reflected to the primary side of the converter.

Bias point is ok

Forward CM - avg - opto TL431.asc

The duty ratio is determined at 37.7% in this simulation. V_{out} is regulated at ≈12 V: bias points are ok.

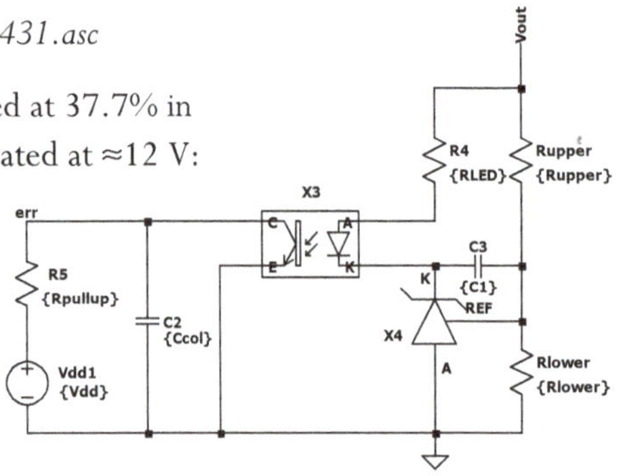

.param Ri=0.48
.param Sn={((N*Vin-Vout)/L)*N*Ri} ← Inductor slope in the primary
.param Smag={Vin*Ri/Lmag} ← Magnetizing slope S_{mag}
.param Sr=0
.param Se=Smag+Sr ← Value passed to the model
.param Mc={1+Se/Sn}

Is S_{mag} enough?

check $m_c > 0.818/(1-D)$

If yes, no need to add an extra ramp, the magnetizing current is enough. If no, determine the amount of S_r to bring Q_p back to unity. There is an example of this approach in the macro found in *forward CM isolated – cbc.asc.*

You can place the CM PWM switch model before or after the transformer. If placed before – as in the circuit – then you must reflect the output inductance with respect to the transformer turns ratio N. As a reminder, N is defined as N_s/N_p in my case, which means 1:0.085 in this example (N_p = 106, N_s = 9 for example). The TL431 is wired in a type 2 compensator where the fast lane is active.

In the forward converter, the magnetizing current plays the role of a free compensation ramp which can help damping the sub-harmonic poles. You need to check if the quality factor Q_p is unity with the magnetizing slope or if a small S_r needs to be added for optimum compensation. A unity Q_p is obtained when m_c is greater than the value given in the left side.

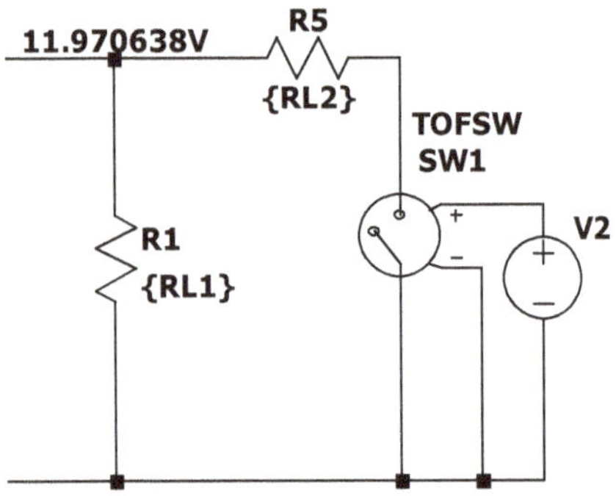

Once the loop is stabilized with adequate margins, you can run a transient step to check that the converter is well stabilized. In this case, the new schematic will include *PWM-CM_TRAN.asc* for the CM averaged model. In this subcircuit, the resonant capacitor has been removed. It avoids a convergence issue in the simulation when the conduction mode changes from CCM to DCM or vice versa.

AC and TRAN Plots of the CM Forward

THE CONVERTER is stabilized with the averaged model for a 2.5-kHz crossover frequency and a 60° phase margin. The Bode plots are shown below for the compensated loop gain, the power stage and the type 2 compensator. The left and right scales for the loop gain, are all centered around zero, which helps to read crossover and phase margin.

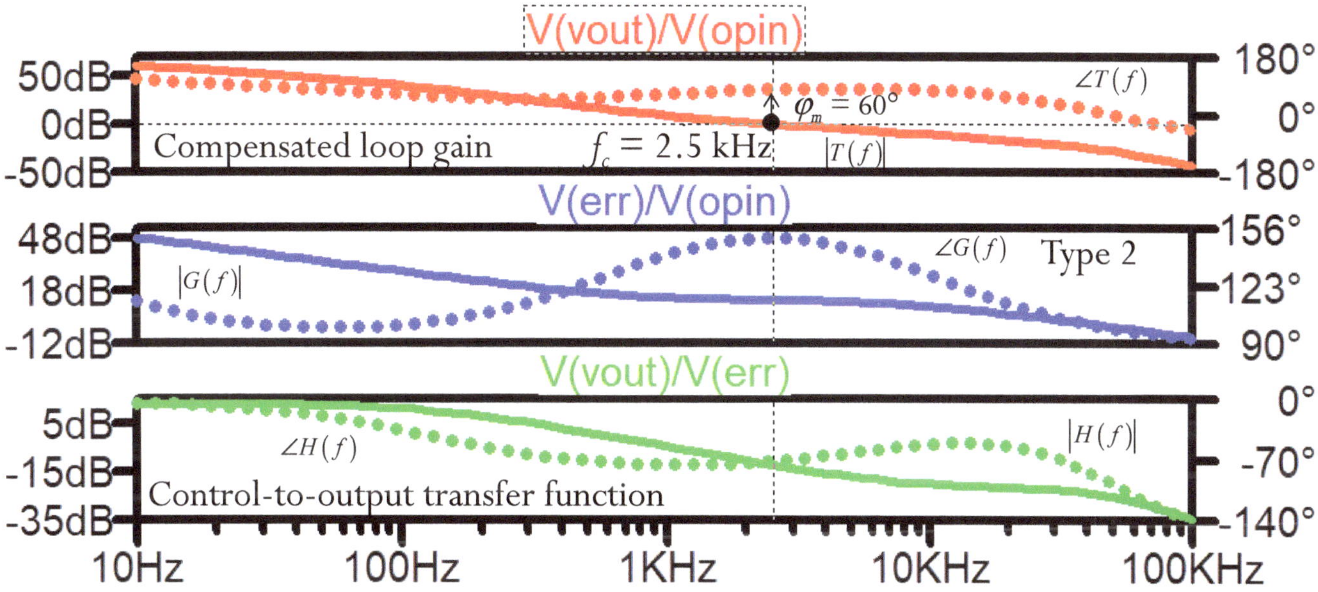

Once the compensated loop gain confirms the converter is stable, you can reflect the magnitude and phase values you have extracted from the control-to-output transfer function at 2.5 kHz, in the transient version of this averaged simulation template. It is easy and fast to change crossover or phase margin, and see the impact on the response.

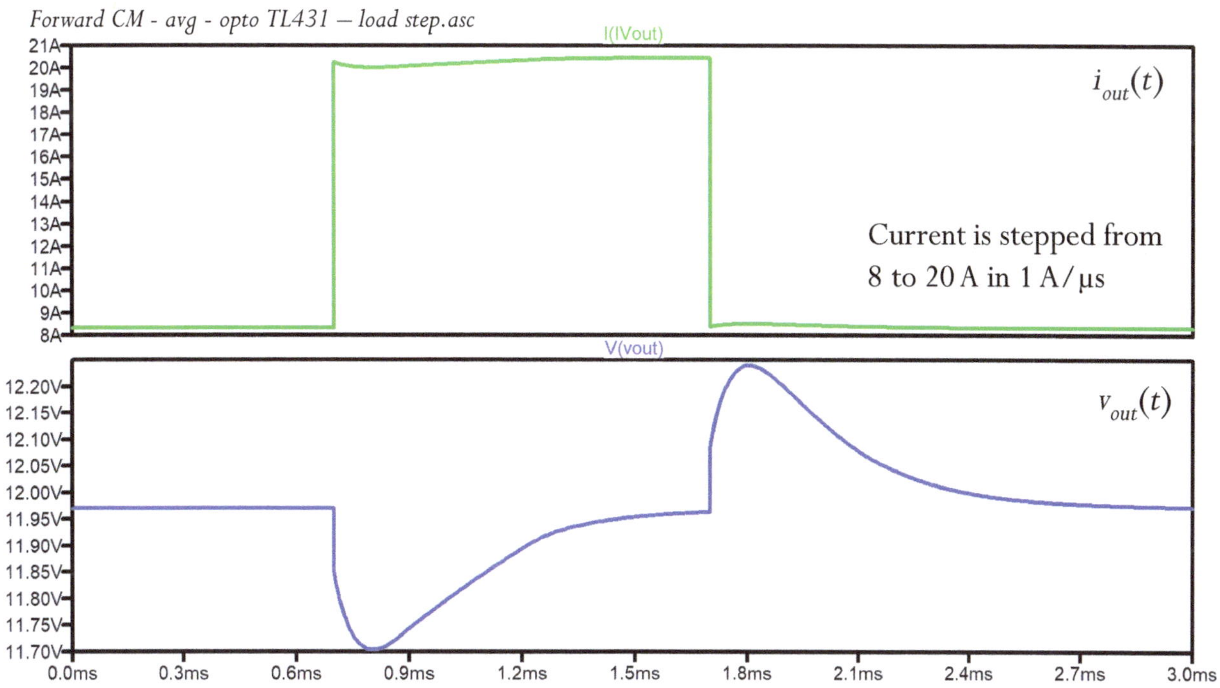

Single-Switch CM Forward Converter

NOW THAT WE HAVE stabilized our converter with the averaged model, we can test a cycle-by-cycle version and verify the stability is correct. You can see a single-switch forward converter to which a demagnetization winding has been added.

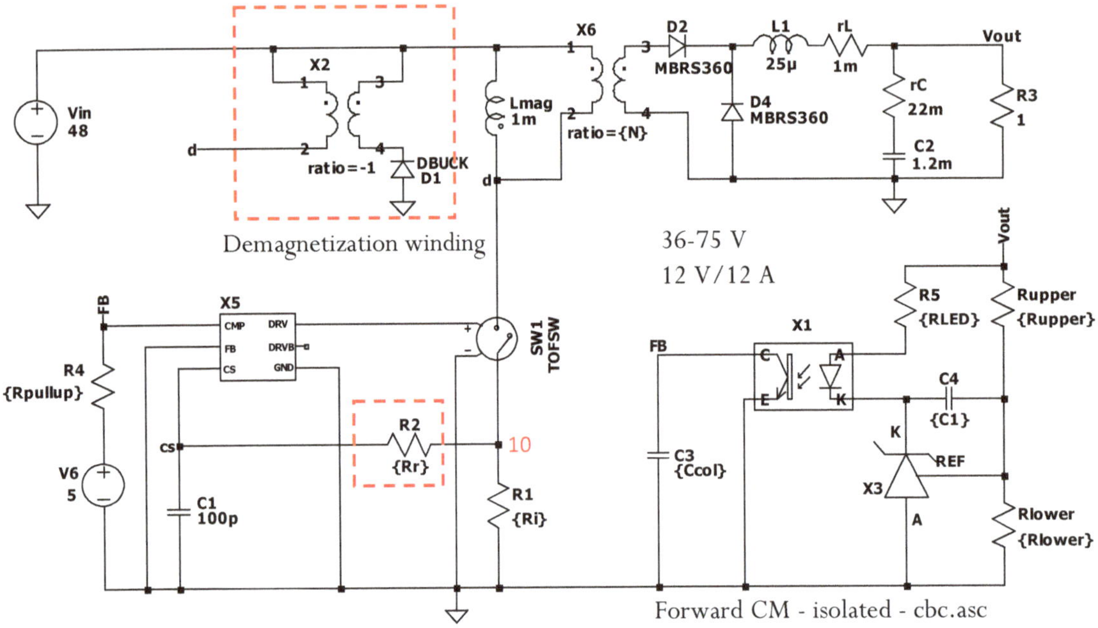

Forward CM - isolated - cbc.asc

A type 2 compensator is built around a TL431. The extra bias resistance across the LED is omitted. Its presence depends on R_4's value and the optocoupler CTR. Crossover has been pushed to 10 kHz.

This winding operates as a flyback converter whose output voltage is V_{in}. When the power switch opens, diode D_1 conducts and applies $-V_{in}$ across L_{mag}, ensuring its demagnetization down to zero. Resistance R_r sets the amount of compensation ramp and its value is calculated by the macro based on the available magnetizing ramp.

```
.param Sramp={2.5/Tsw}
.param D={(Vout/(N*Vin))*1.2} ;
.param Smag={Vin*Ri/Lmag}
.param Sn={(((N*Vin-Vout)/L)*N)*Ri}
.param Se={Sn*(0.818/(1-D)-1)}
.param S1=If(Smag>Se,0,Se-Smag)
.param Rr={(S1/Sramp)*19k+100}
```

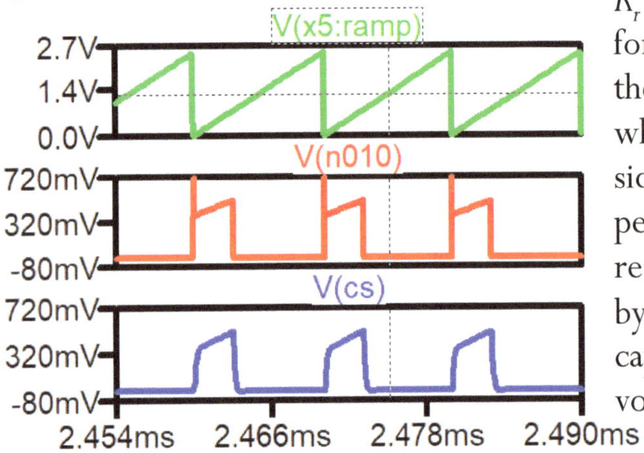

Parameter S_e is evaluated to determine the amount of needed ramp to reduce the quality factor of the sub-harmonic poles Q_p to unity. Then S_1 verifies if the magnetizing current is high enough to meet the S_e requirement. If yes, then there is no need for more ramp and R_r is set to 100 Ω: it creates a small *RC* filter for the current sense voltage. If S_{mag} is too low, then S_1 is adjusted as the difference between what is available and what is needed. In the left side, you can see the internal ramp (2.5 V peak), the raw voltage across the sense resistance and, at the bottom, the voltage seen by the CS pin. As R_r = 1 kΩ with a 100-pF capacitor, you can observe a clean $v_{CS}(t)$ voltage.

Transient Waveforms

NOW THAT THE CONVERTER is stabilized, we can check the start-up sequence and confirm the magnetizing current remains discontinuous at all times. The loop smoothly closes when V_{out} reaches 12 V, without a noticeable overshoot. The magnetizing current remains below 150 mA peak for a 48-V input voltage.

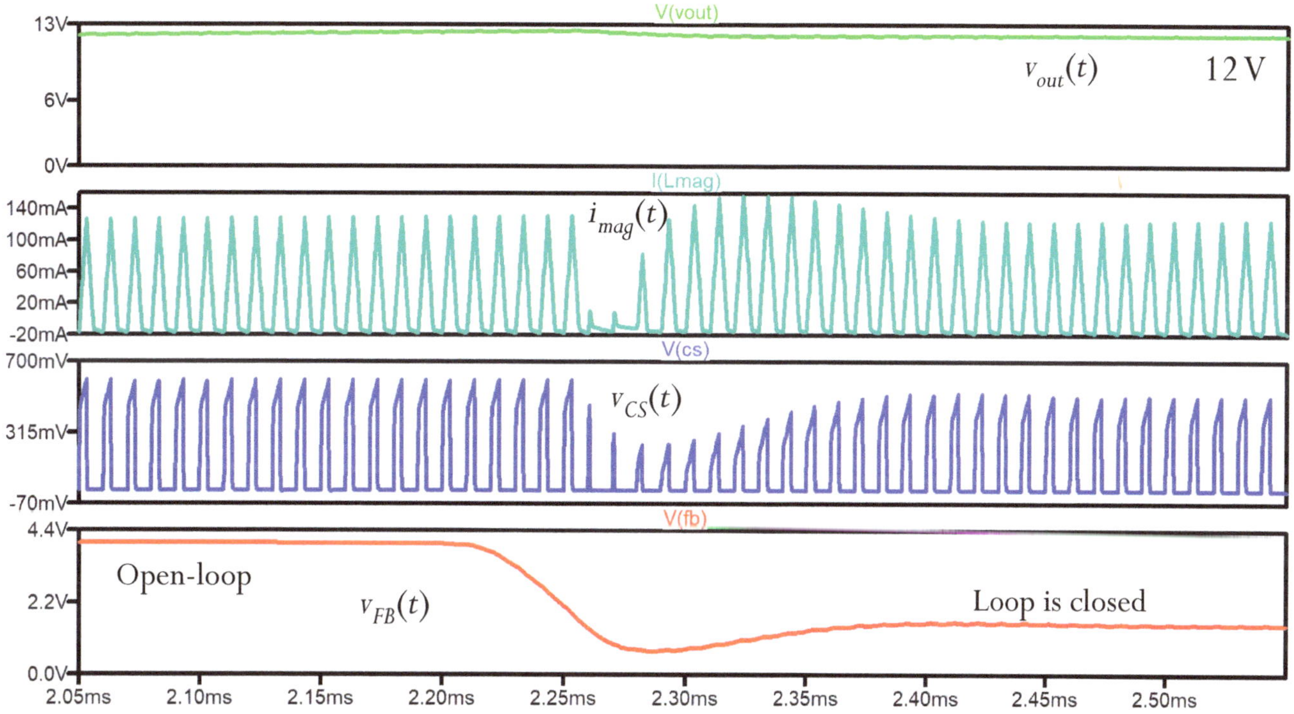

To speed up the simulation time and soften convergence issues, I often use the following set of options, well suited for power conversion:

.options abstol=1u vntol=1m reltol=0.01 gmin=100p method=gear

Sometimes these parameters are too relaxed and you may need to tweak them. Also, I often limit the maximum time step to 50-100 ns when I see waveforms lacking granularity.

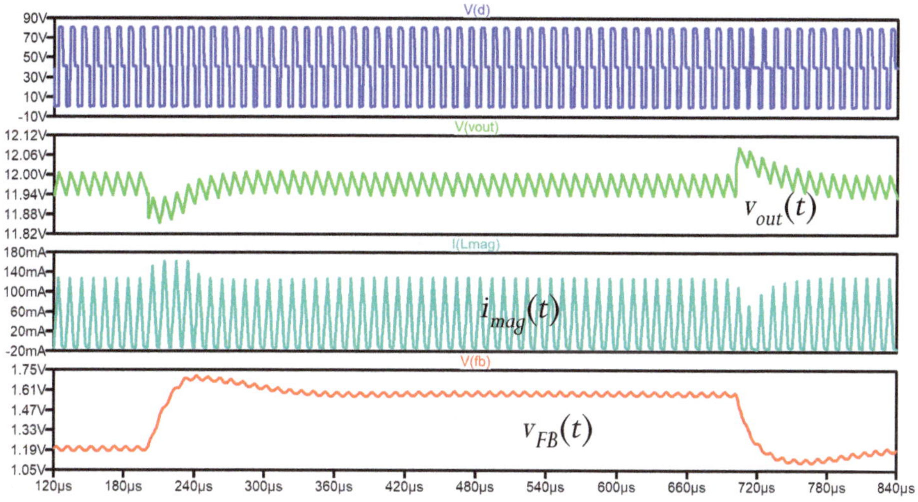

The transient response is excellent and you can see the drain voltage reaching twice V_{in}. Adding some leakage inductance to the transformer, will generate unavoidable peaks, which must remain below the MOSFET BV_{DSS}.

Two-Switch Forward with UC3845

AS POWER INCREASES, the 2-switch forward converter offers a reliable and robust structure for levels up to 700-800 W, a typical range for the so-called *silver boxes* found in PCs and servers. The demagnetization winding has disappeared and two diodes now clamp the primary side to the input voltage, ensuring a complete demagnetization cycle. The leakage inductance is included and affects the voltage stress of the MOSFETs.

I built the artificial ramp generator from the low-impedance DRV output. The oscillator is untouched, which is better for the noise immunity. The macro automates all components calculations.

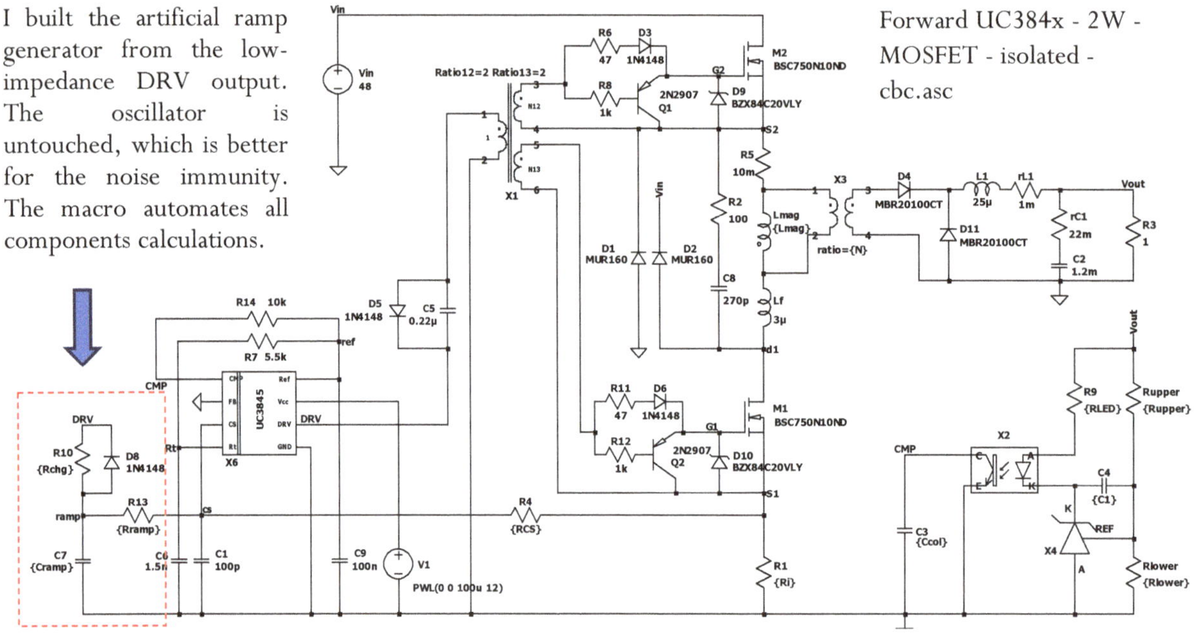

Forward UC384x - 2W - MOSFET - isolated - cbc.asc

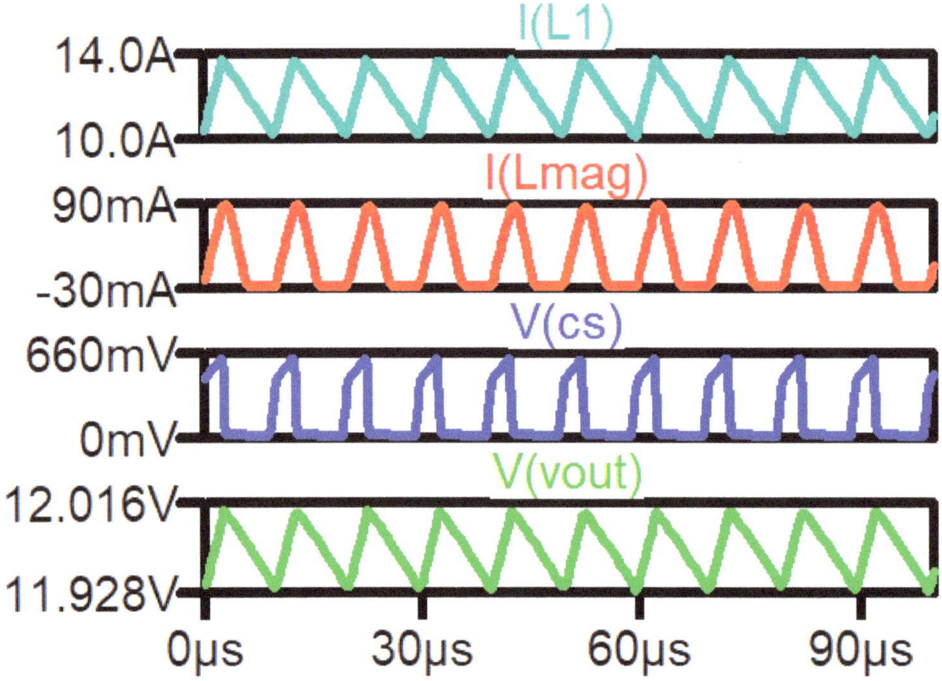

The converter delivers 12 V/12 A. The magnetizing current correctly returns to zero ampere. The pause you see on top of the red curve, is the overlap between the secondary diodes, when one takes over the other one.

The MOSFET transistors are driven from a two-output transformer, as a bootstrapped driver would be difficult to implement considering the refueling of the bootstrap capacitor. The load step confirms the loop is stable (*Forward UC384x - 2W - MOSFET - isolated - cbc - loadstep.asc*).

Multi-Output Forward Converter

PERSONAL COMPUTER silver boxes have extensively used the forward in multi-output applications, delivering different voltages like 3.3 V, 5 V and 12 V, with various current capabilities. To improve the regulation between the outputs, it was common to couple the inductors which offered an excellent transient response on outputs not included in the regulation loop. It is possible to reproduce a similar circuit, here with a regulated 12 V featuring an auxiliary 5-V output. The two inductors are coupled via my generic transformer model, but a coupling factor could also be used, especially if you plan on adding output windings. A UC3845 is used for controlling this single-switch converter.

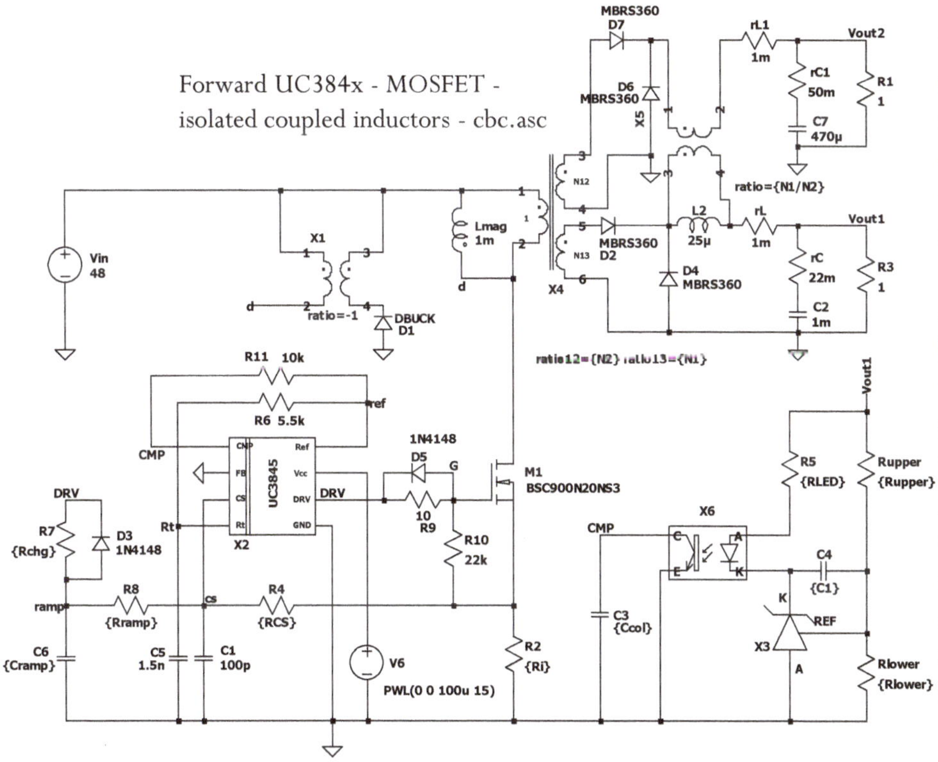

The main 12-V output is regulated by the TL431, but the 5 V is left unregulated. I could have installed a weighted feedback, but did not for simplicity. The UC3845 drives a single switch and more windings can be added to the main transformer if needed. These silver boxes were also implementing magnetic post-regulators for improved transient performance. An averaged model - *Forward CM - avg - TL431 - coupled inductors.asc* – is also available for ac analysis.

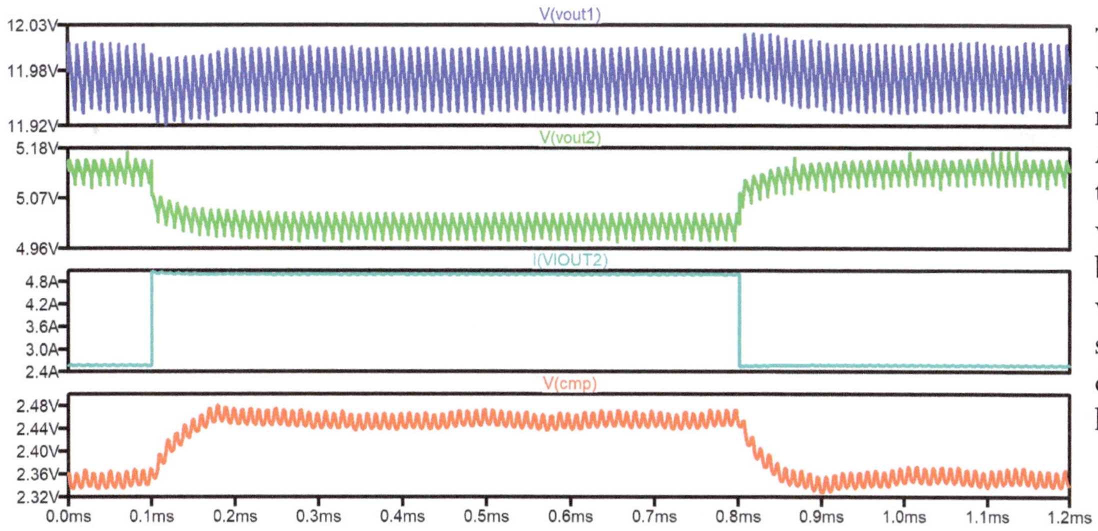

The main 12-V output does not see the 3-A transient on the 5-V rail which behaves very well, with a small drop during the load step.

Home

Active-Clamp Forward

THE ACTIVE-CLAMP forward converter is popular in high-density dc-dc bricks in the telecom market, for instance. I remember designing a 3.3-V/30-A converter using the NCP1565/66 from *onsemi*, years ago. Below is a simplified non-isolated version of it, but still implementing self-driven synchronous rectifiers in the secondary side. A current-sense transformer conveys the primary current to the controller. The control of the low-side P-channel transistor is done by inverting the DRVB output. This reset scheme could also be a high-side type (with a N-channel) but for a low voltage application, the P-channel is simple and efficient.

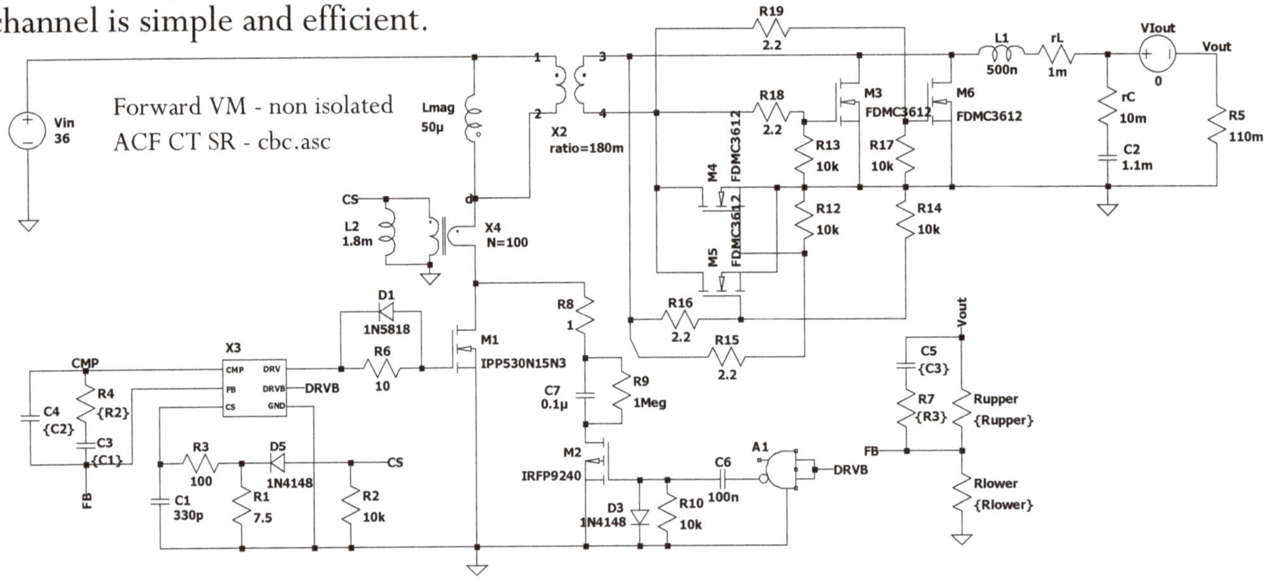

The magnetizing current shows that the transformer works in quadrants I and III and can accept a duty ratio greater than 50%, naturally allowing a wider input voltage range than with a classical single-switch structure. Switching losses are also reduced owing to quasi-ZVS operation.

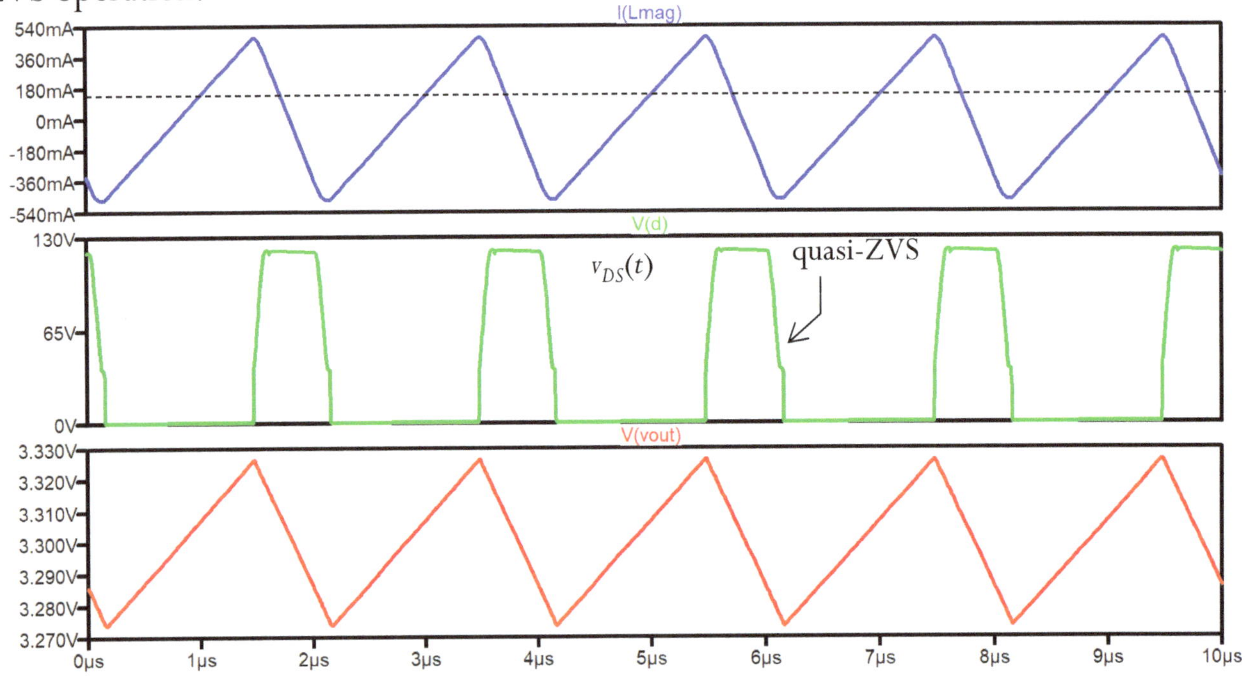

Full-Bridge CM Converter

THE FULL-BRIDGE structure is well suited for powering high-power converters. The below circuit illustrates a 5-V/100-A power supply. The schematic diagram is obviously simplified, but still offers a convenient way to test the transformer turns ratio, confirm the magnetizing current is well within safe limits, or look at the various stresses in the semiconductors and the passive elements. The control is made via the generic current-mode model and drives the four transistors in diagonal: SW3/SW5 then SW4/SW2. The input current is monitored via the 0-V dummy source V_9, but I could also have used a current transformer.

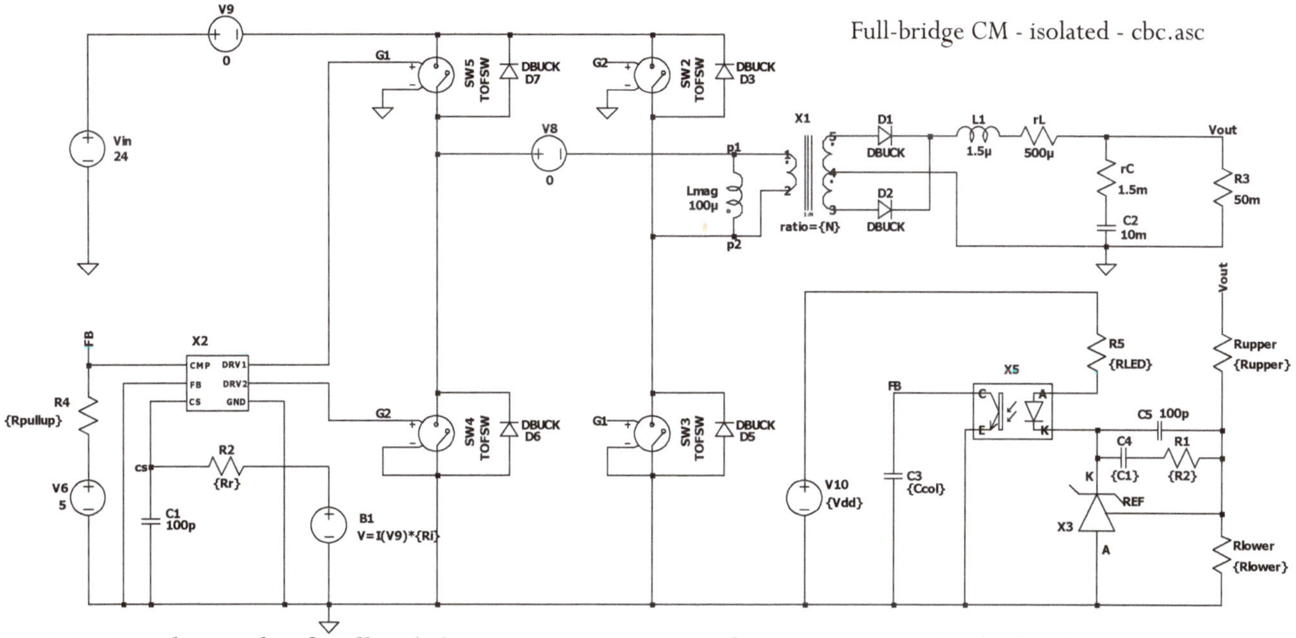

A TL431 closes the feedback loop in a type 2 configuration in which the fast lane has been disabled. C_5 limits the high-frequency noise but the second pole is provided by C_3, together with R_{pullup}. As expected, the power transformer operates in quadrants I and III.

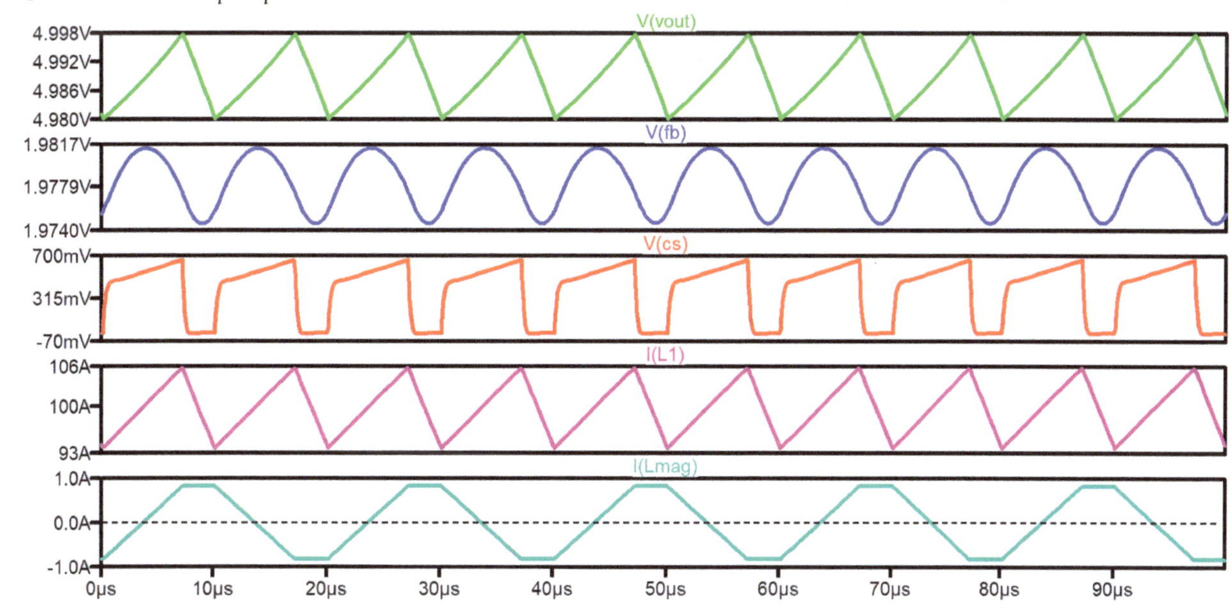

Full-Bridge Phase-Shifted VM Converter

WE ARE NOW ENTERING the high-power arena, with converters capable to deliver kilowatts of power. The phase-shifted full-bridge converter belongs to this category. As shown in the diagram below, the four power switches are actuated as two half-bridges in which a dead time favors zero-volt switching (ZVS) in certain conditions. The structure can be operated in voltage- or current-mode control. In VM, the control-to-output ac response does not exhibit the peaking found in a buck-derived converter. Damping is partly provided by the leakage inductance, which also ensures ZVS.

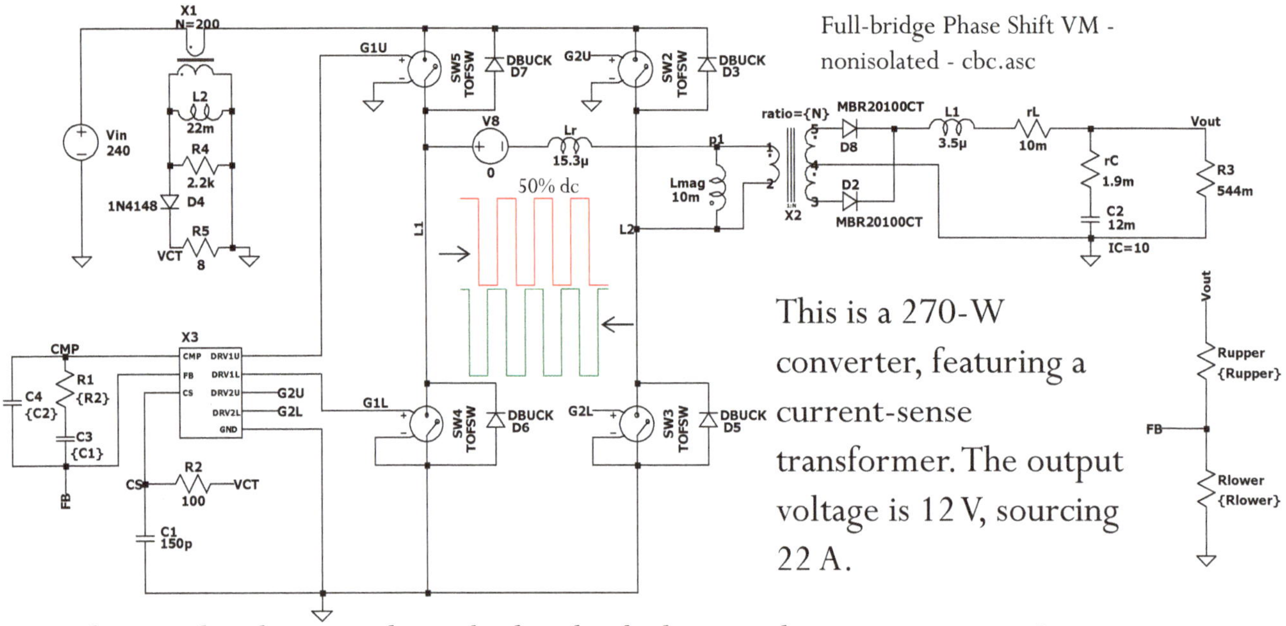

This is a 270-W converter, featuring a current-sense transformer. The output voltage is 12 V, sourcing 22 A.

Considering the damping brought by the leakage inductance, a type 2 compensator is used to stabilize the converter. The primary-side current shape in L_r is typical of that type of phase-shifted converter.

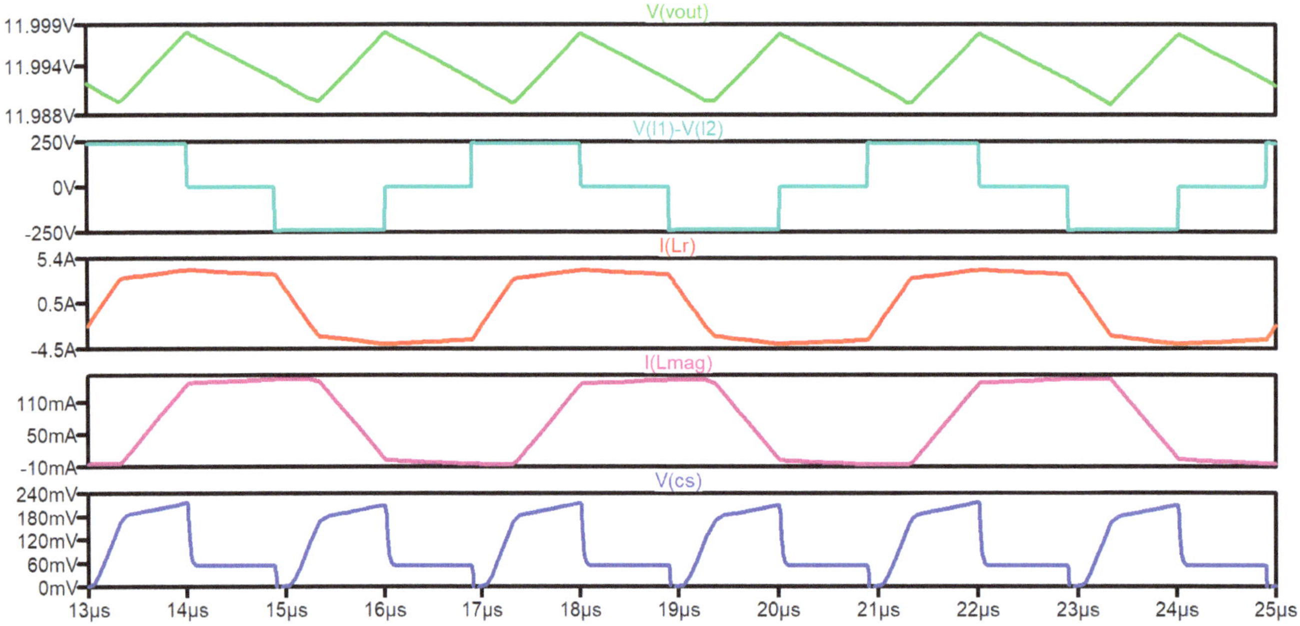

Dual-Active-Bridge Converter

THE-DUAL-ACTIVE-BRIDGE converter implements two full bridges located in the primary and the secondary sides. It is a bidirectional converter by construction. A dc-block capacitor is often needed in the primary side to prevent transformer saturation. For the same reason, a second capacitor takes place in the secondary, in case you design a bidirectional converter. However, passive and active techniques exist to prevent flux walkaway and thus suppress these bulky (and expensive) capacitors.

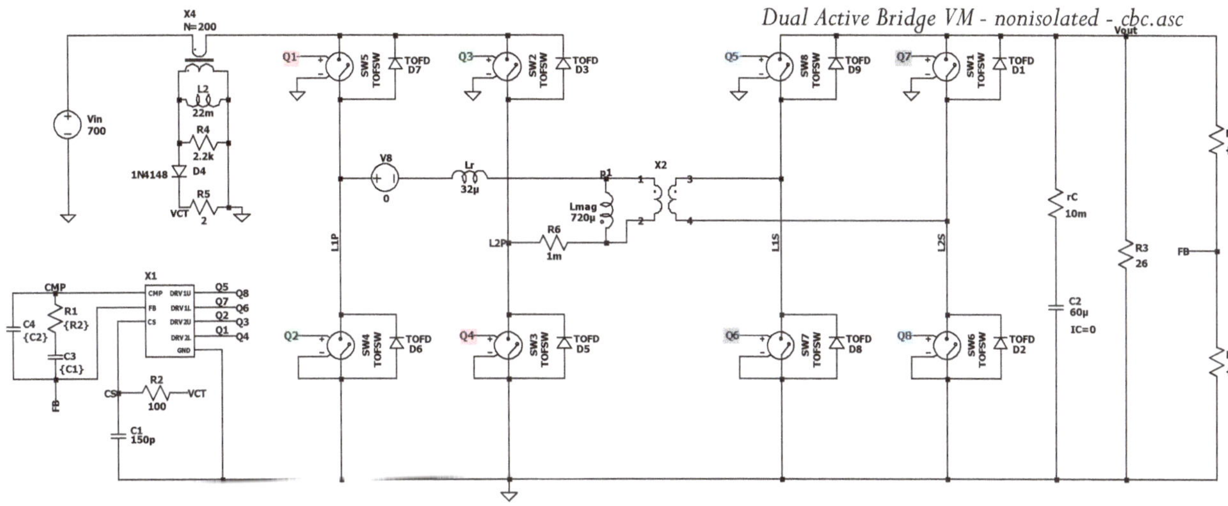

The labeling of the gates shows you how the switches are driven. This is a single phase-shift type of control, where the series inductance L_r controls the power flow between the primary and secondary sides. The steady-state waveforms show a 500-V output for a 10-kW output power. The compensation is made via a type 2, considering the 1st-order response of the converter. As mentioned, in a real project, a series capacitor or an extra loop monitoring the magnetizing current is necessary for a reliable operation.

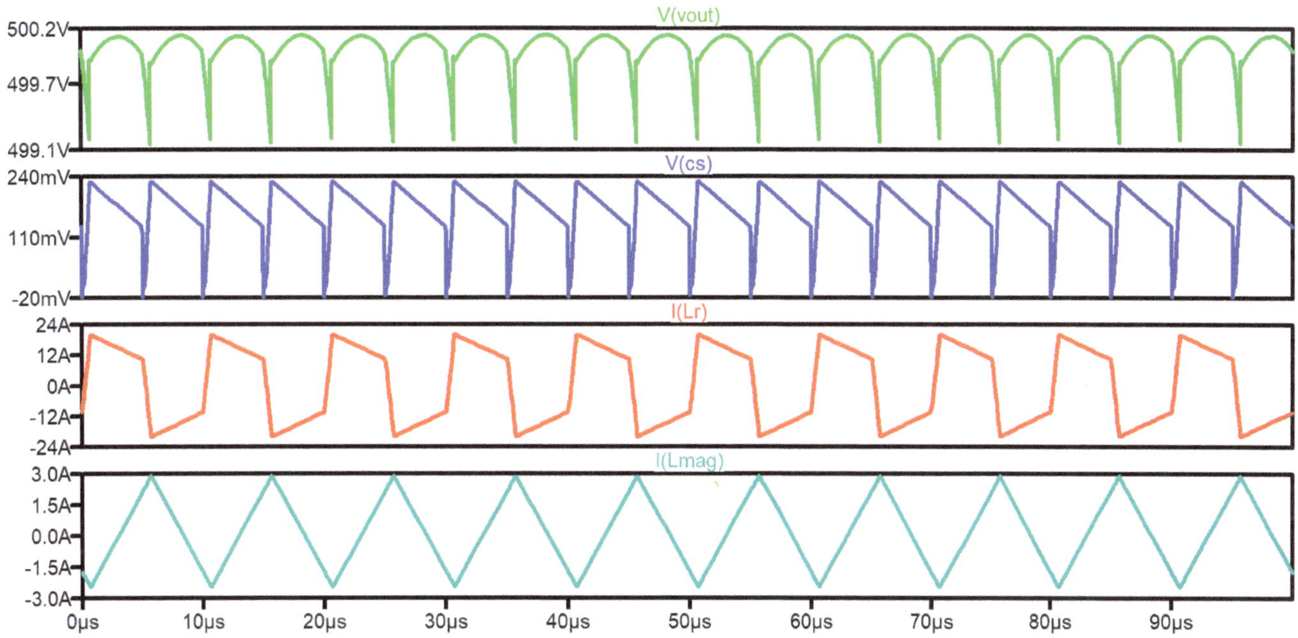

The Boost Converter – VM

THE BOOST CONVERTER is part of the basic switching cells, together with the buck and buck-boost converters. It can be operated in voltage- or current-mode control, usually at a fixed switching frequency, but boundary-mode is also very common in power factor correction (PFC) applications. We have different ways to simulate a boost converter, but the fastest way – to test the operating point and the ac response – remains the averaged model. Keep in mind that, considering the absence of switching component in this type of circuit, the simulation time is truly flashing. It is thus a good thing to start with this approach, especially for closing the loop before considering a cycle-by-cycle version.

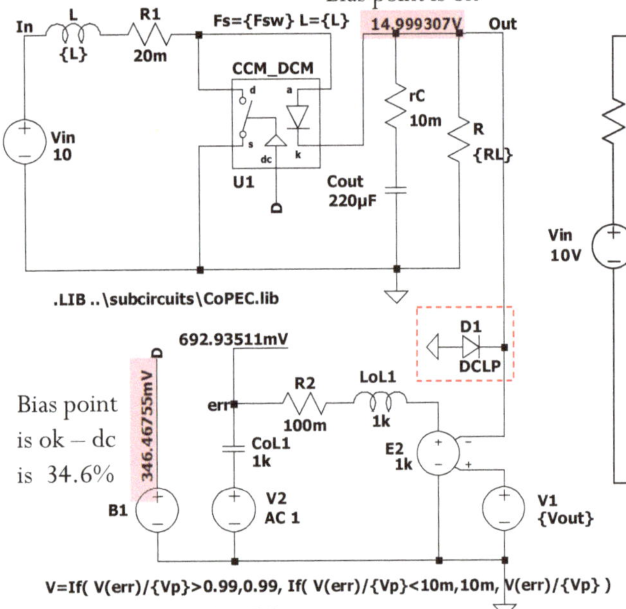

Boost VM - avg - CoPEC.asc

A clamp diode (D_1) is added with the CoPEC model to improve dc convergence.

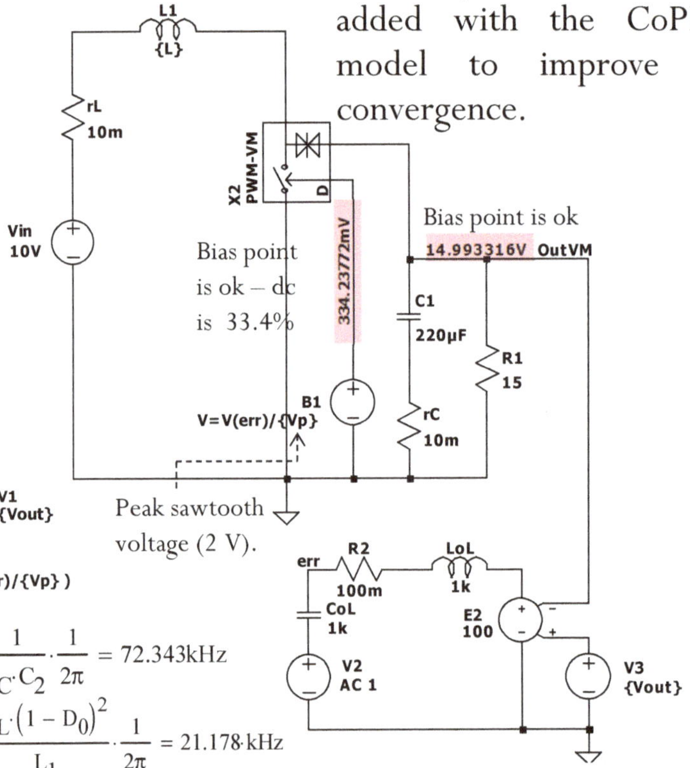

$$f_{z1} := \frac{1}{r_C \cdot C_2}\cdot\frac{1}{2\pi} = 72.343\text{kHz}$$

$$f_{z2} := \frac{R_L\cdot(1-D_0)^2}{L_1}\cdot\frac{1}{2\pi} = 21.178\cdot\text{kHz}$$

The CoPEC model lends itself well to simulating a boost converter. No problem either with the VM-PWM switch model, as both subcircuits automatically toggle between DCM and CCM operation. The CCM ac response appears in the right side. It confirms the presence of the RHP zero occurring before the ESR zero in this example.

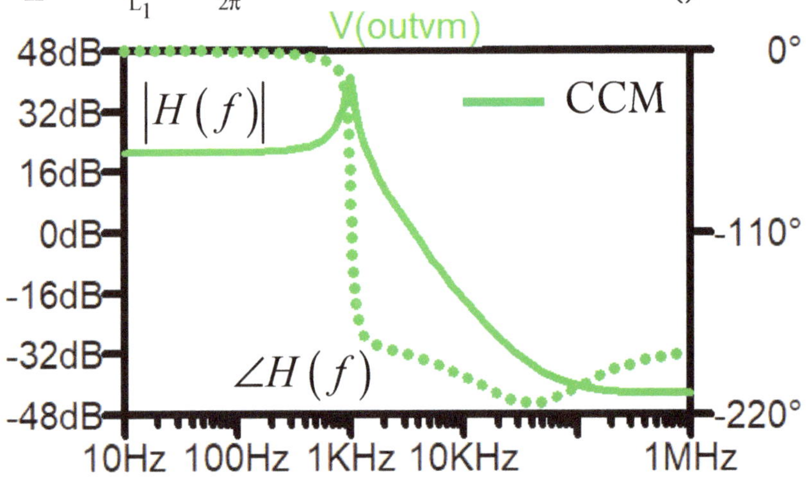

The Boost Converter – CM

IN THE PWM SWITCH MODEL, originally derived with a buck converter, the current *leaves* terminal *c*. However, in the boost configuration, the current will *enter* terminal *c* as shown below. That is the reason why the sense resistance is negative in this particular application. As usual, make sure the operating bias points are correct – the boost converter delivers 15 V from a 10-V source – before considering the ac results. The duty ratio is 33.6% for a 5-Ω load while the control voltage at node V_c is 1.39 V.

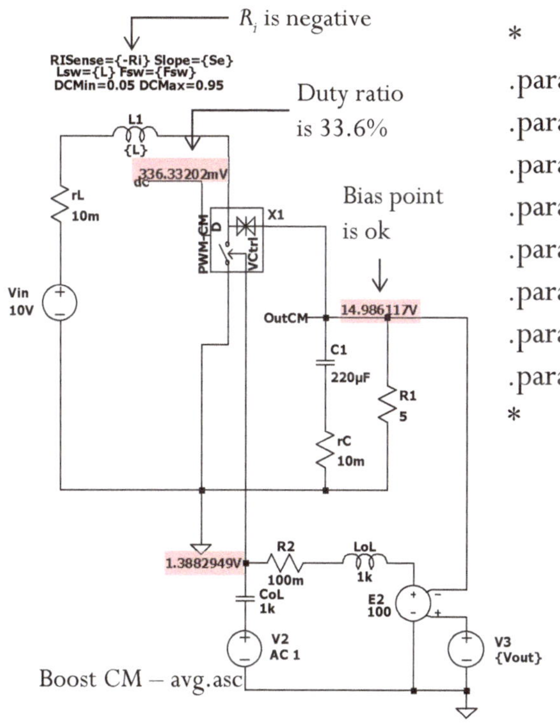


```
*
.param L=50u ; inductance value
.param Vin=10 ; input voltage
.param Vout=15 ; output voltage
.param Fsw=300k ; switching frequency
.param Ri=0.3 ; sense resistance (neg. for a boost)
.param Sn={Vin*Ri/L} ; on-time inductor current slope
.param Mc=1.0 ; amount of slope compensation
.param Se={(mc-1)*Sn} ; external ramp level for meeting m_c
*
```

The parameters to be passed are shown above. Inductance and frequency are necessary to calculate the location of the sub-harmonic poles, but also for determining the critical condition below which the part toggles between CCM and DCM.

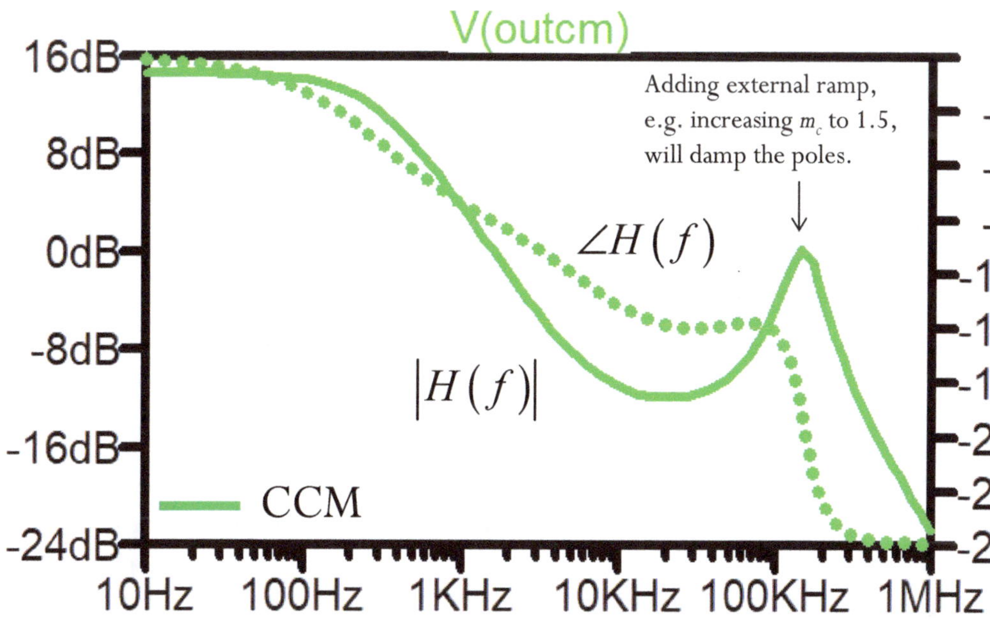

The boost converter operates in CCM in this example. You can clearly see the peaking incurred to the double poles located at $F_{sw}/2$.

A Simple Boost VM with FRA

RATHER THAN using an averaged model, we can take advantage of the built-in frequency-response analyzer. For this, we first assemble a simple cycle-by-cycle converter featuring a pulse-width modulator. The ramp peak amplitude is 2 V which, together with a 680-mV dc bias, gives a duty ratio of 34%. The output voltage is therefore $V_{in}/(1\text{-}0.34)$ or ≈15 V.

In an open-loop configuration, the modulation amplitude is constant and easy to adjust. I have selected 20 mV here.

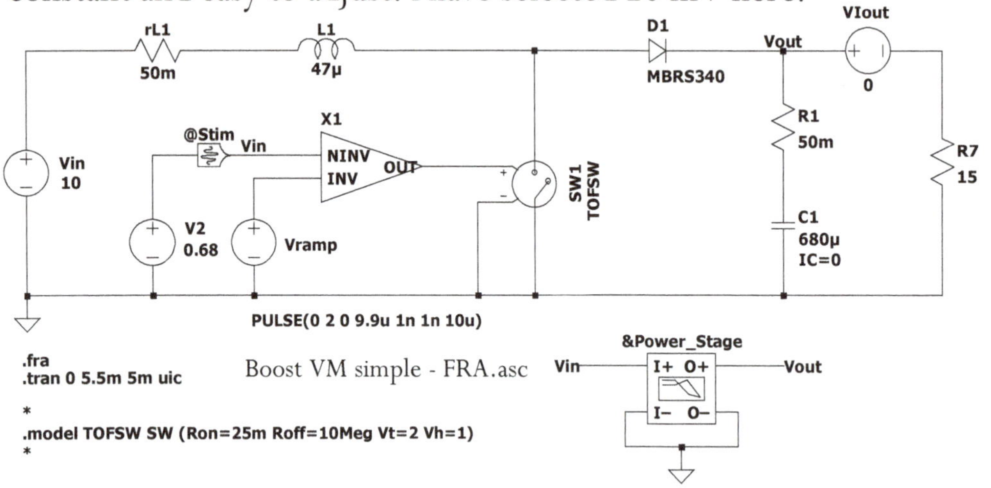

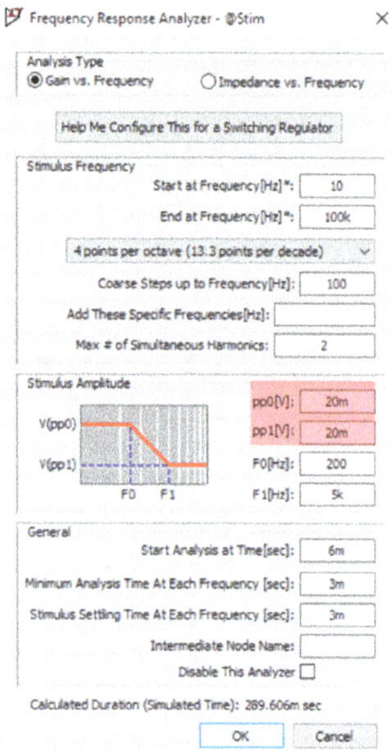

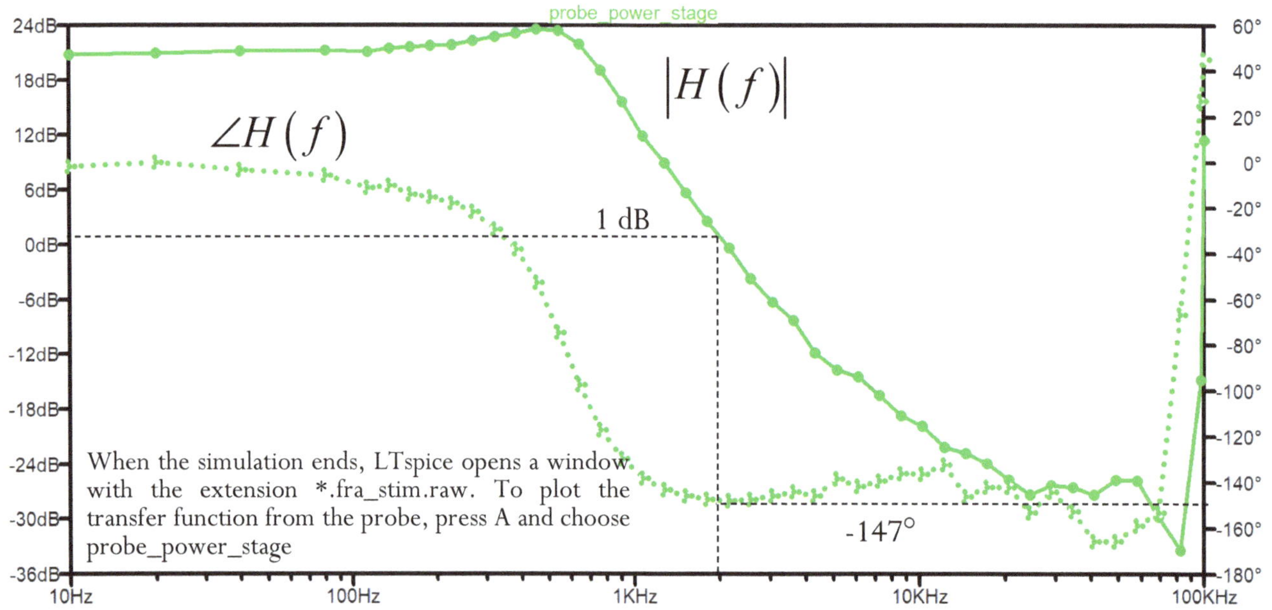

The crossover frequency is selected away from the resonant peak, at 2 kHz. The RHP zero is located at ≈22 kHz so we are below 20% of this worst-case value. Just read the magnitude and phase values at 2 kHz and enter them in the macro of the cycle-by-cycle circuit. The compensator will automatically be calculated to force crossover at this point, with the desired phase margin. Be careful not push f_c too far or the RHPZ will manifest its presence.

Compensated Averaged Boost Converter

AN AVERAGED MODEL lends itself very well to compensating a switching cell like the boost converter. In voltage-mode, the PWM switch is easy to use and the schematic is quickly assembled to fit the proper configuration. With this template, you can obtain the control-to-output transfer function of the power stage, but also the Bode plot of the compensator. The final probing will give you the compensated gain T, from which you will extract the crossover frequency f_c, the phase margin PM or φ_m and the gain margin GM.

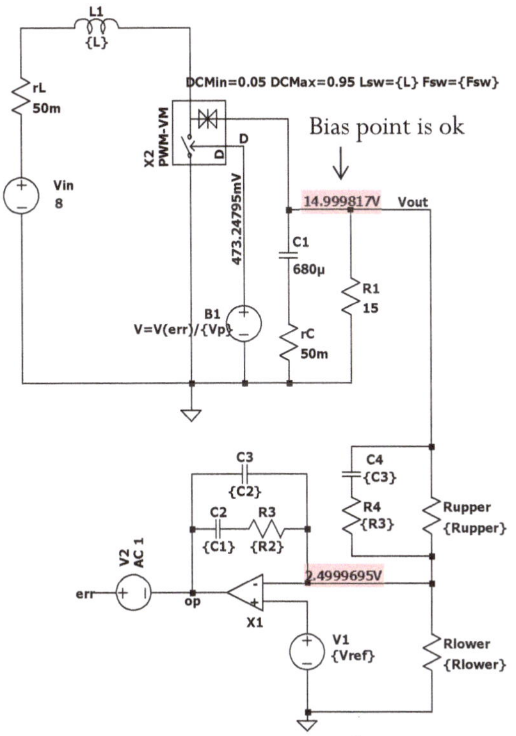

Boost VM - avg - compensated.asc

```
*
.param Gfc=-1 ; magnitude at crossover extracted from power stage response
.param PS=-153 ; phase lag at crossover extracted from power stage response
.param fc=2k ; targeted crossover – your design goal
.param PM=60 ; choose phase margin at crossover – your design goal
*
.param fz1=300 ; one zero is placed slightly below resonance
.param fz2=400 ; one zero is placed at resonance
.param fp2=50k ; this pole is usually placed at Fsw/2
*
.param Ibias=1m ; the bias current for the resistive divider
.param Vref=2.5 ; the selected reference voltage
*
.param Rlower={Vref/Ibias} ; the upper resistance value
.param Rupper={(Vout-Vref)/Ibias} ; the low-side resistance value
.param boost={PM-PS-90} ; determine how much phase boost is needed
.param G={10**(-Gfc/20)} ; check the needed gain at fc
.param fp1={fc/tan(atan(fc/fz1)+atan(fc/fz2)-atan(fc/fp2)-boost*pi/180)}
.param a={sqrt((fc**2/fp1**2)+1)}
.param b={sqrt((fc**2/fp2**2)+1)}
.param c={sqrt((fz1**2/fc**2)+1)}
.param d={sqrt((fc**2/fz2**2)+1)}
.param R2={((a*b/(c*d))/(fp1-fz1))*Rupper*G*fp1}
.param C1={1/(2*pi*fz1*R2)}
.param C2={C1/(C1*R2*2*pi*fp1-1)}
.param C3={(fp2-fz2)/(2*pi*Rupper*fp2*fz2)}
.param R3={Rupper*fz2/(fp2-fz2)}
```

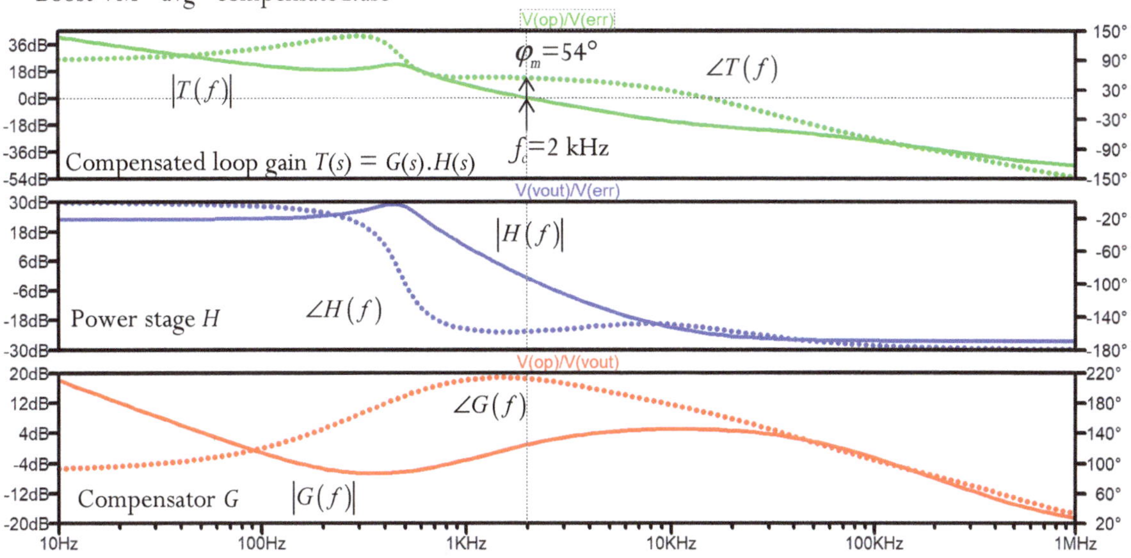

The simulation confirms that our goals of crossover and phase margins are met. DCM should also be explored to verify the margins in this mode.

The frequency response is obtained in less than 1 s and it gives you the full picture, with the power stage response, the compensator and the compensated loop gain.

Monte-Carlo Analysis

ONCE THE CONVERTER is stabilized, you must explore the effects of components variability on crossover frequency and phase/gain margins [7]. This is the only way to guarantee that stability won't be at stake after the production has started or after several years of operation. In this page, I have illustrated how to assign tolerances to resistors and capacitors, then run a certain numbers of simulations. Averaged modeling offers the fastest way to cover multiple tolerance scenarios. The lack of switching component makes each analysis extremely fast and you can run tens of sweeps in a very short time. Cycle-by-cycle simulations with the FRA symbol cannot compete.

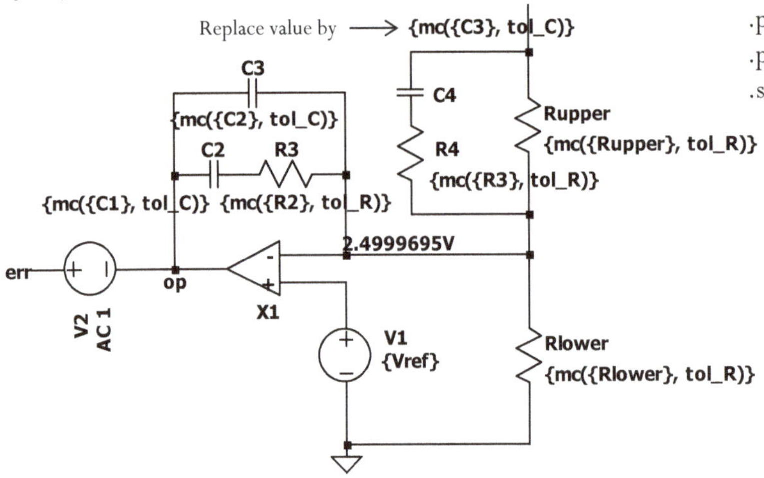


```
.param tol_C=.2 ; +/- 20% component tolerance
.param tol_R=.01 ; +/- 1% component tolerance
.step param X 0 30 1 ; cycle Monte Carlo runs
```

Boost VM - avg - compensated - Monte Carlo.asc

Tolerances are assigned and I have chosen a uniform distribution law to sweep the values. Gaussian distribution is available for non-sorted components. Once the simulation is over, all curves are conveniently aggregated on a single graph, as shown below.

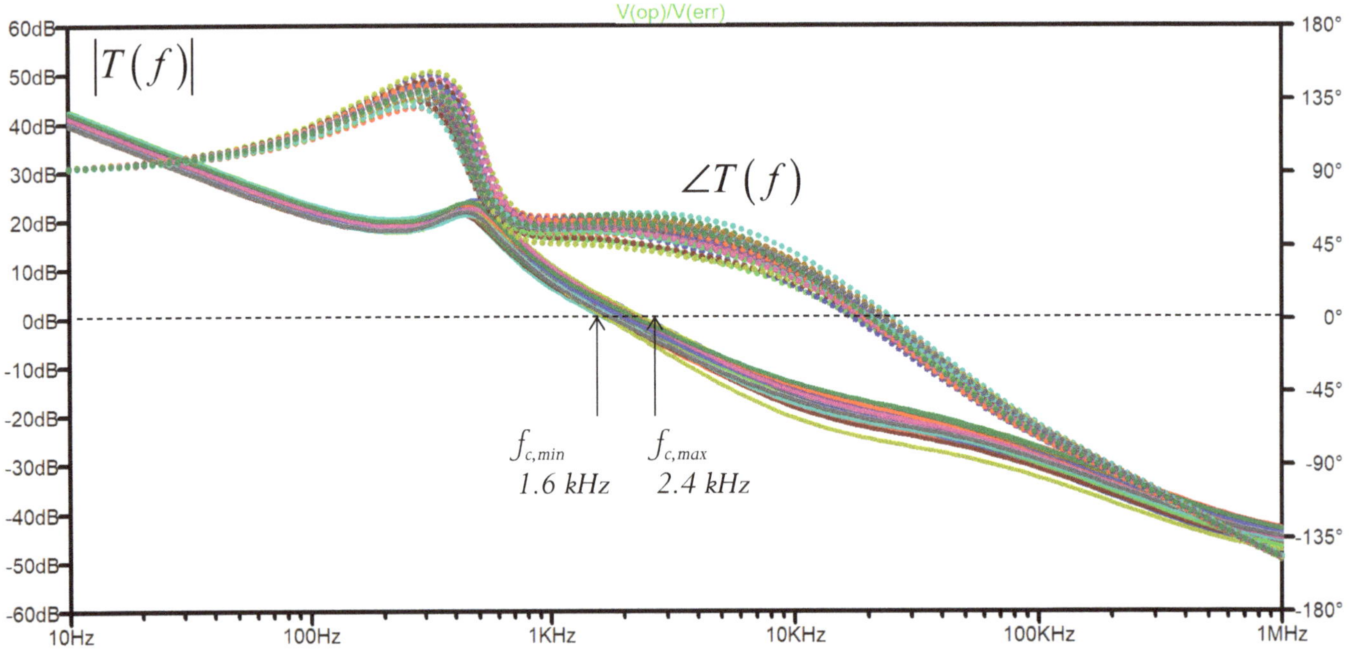

When analyzing the data, it appears that the lowest phase margin (44.5°) occurs with a crossover frequency of 1.7 kHz. On the opposite, the highest crossover is 2.4 kHz with a 53° phase margin. Gain margin is 18 dB as a minimum, which is excellent.

Analyzing the Data

AFTER THE MONTE CARLO sweep, it is important to analyze the obtained data. LTspice lets you program a simple measurement routine with the .MEAS or .MEASURE keywords placed in a control block. To extract crossover, phase and gain margins, these are options to print these data in the .LOG file:

```
.meas ac crossover_freq when mag(v(op)/V(err))=1
.meas ac phase_margin find v(op)/V(err) when mag(v(op)/V(err))=1
.meas ac gain_margin find v(op)/V(err) when ph(v(op)/V(err))=0
```

The first statement indicates that when the loop gain magnitude – v(op)/V(err) in this example – reaches 1 or 0 dB, the frequency is recorded. The loop phase is then sampled at the same position. The gain margin is the loop gain recorded at the point where the phase reaches 0°. These results are available in the .LOG file after the simulation has run.

```
crossover_freq: mag(v(op)/V(err))=1 AT 1995.19300503
phase_margin: v(op)/V(err)=(0dB,54.5675845615°) at 1995.19300503
gain_margin: v(op)/V(err)=(-20.0183585432dB,6.14828831369e-16°) at 20428.9776009
```

If crossover can be easily extracted for exporting to Excel, the phase and gain margins are complex numbers while a real value would be more convenient to treat. You will thus need a specific routine in Excel to extract these margins and present them in a format that is compatible with graph generation. The keyword *.options meascplxfmt=cartesian* lets you turn the phase results into real and imaginary numbers, perhaps easier to process with the ATAN2 function in Excel. Below are results for f_c and the phase margin.

Measurement: crossover_freq

step	mag(v(op)/V(err))=1
1	2112.95261173
2	2411.07901648
3	1754.05789498
4	1658.48064658
5	2051.61317194
6	1951.52832292

Measurement: phase_margin

step	v(op)/V(err)	at
1	(0dB,59.3277109512°)	2112.95261173
2	(0dB,52.9731651681°)	2411.07901648
3	(0dB,51.8081909748°)	1754.05789498
4	(0dB,63.3735892305°)	1658.48064658
5	(0dB,53.7517377453°)	2051.61317194
6	(0dB,61.1079732533°)	1951.52832292

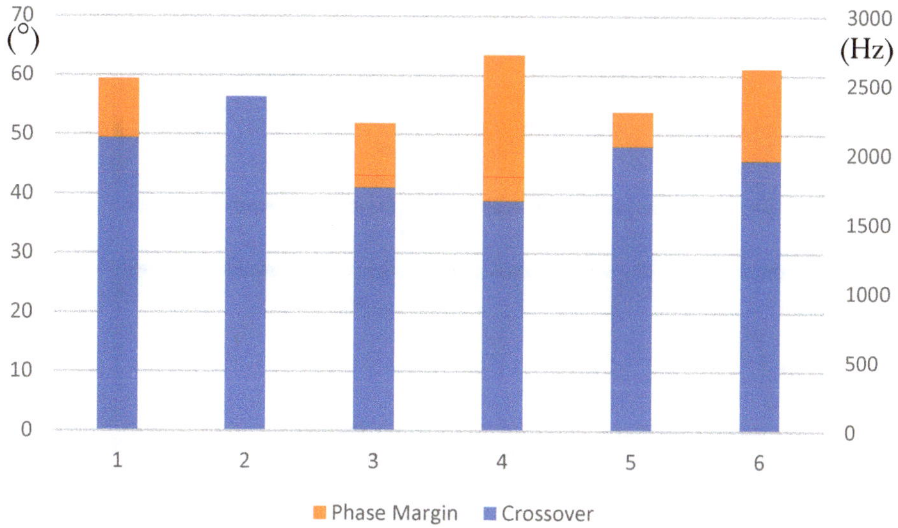

I have imported into Excel crossover frequency and phase margin data for this 6-sweep Monte Carlo run. There are plenty of ways to plot these results, the left-side graph is a possible example of what can be done.

A Cycle-by-Cycle VM Boost Converter

NOW THAT THE boost converter has been stabilized, we can run a cycle-by-cycle simulation and check the compensation network obtained with the averaged model. I can use the generic model which features a complementary output, well suited for implementing synchronous rectification. Considering the high-frequency switching noise, we can observe the output voltage via a simple R_7C_6 network which filters out all the spurious pulses for a clearer waveform.

The simulation time is short and I had results in less than 10 s on my machine. The inductor current confirms the CCM operation. The red curve represents the current-sense information probed after the R_4C_3 network, which efficiently filters out unwanted switching spikes.

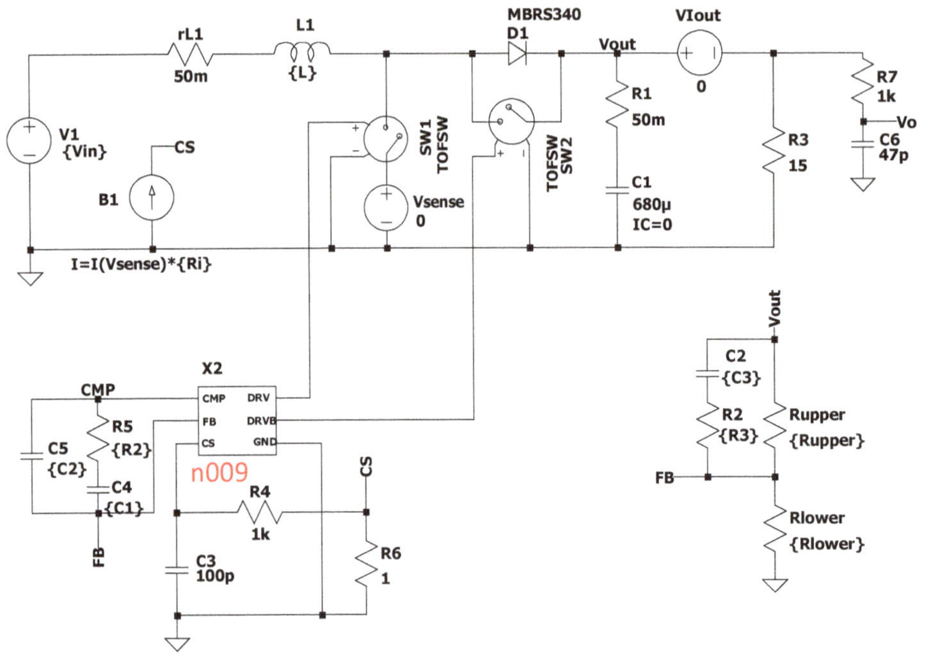

Boost Synchro VM - cbc.asc

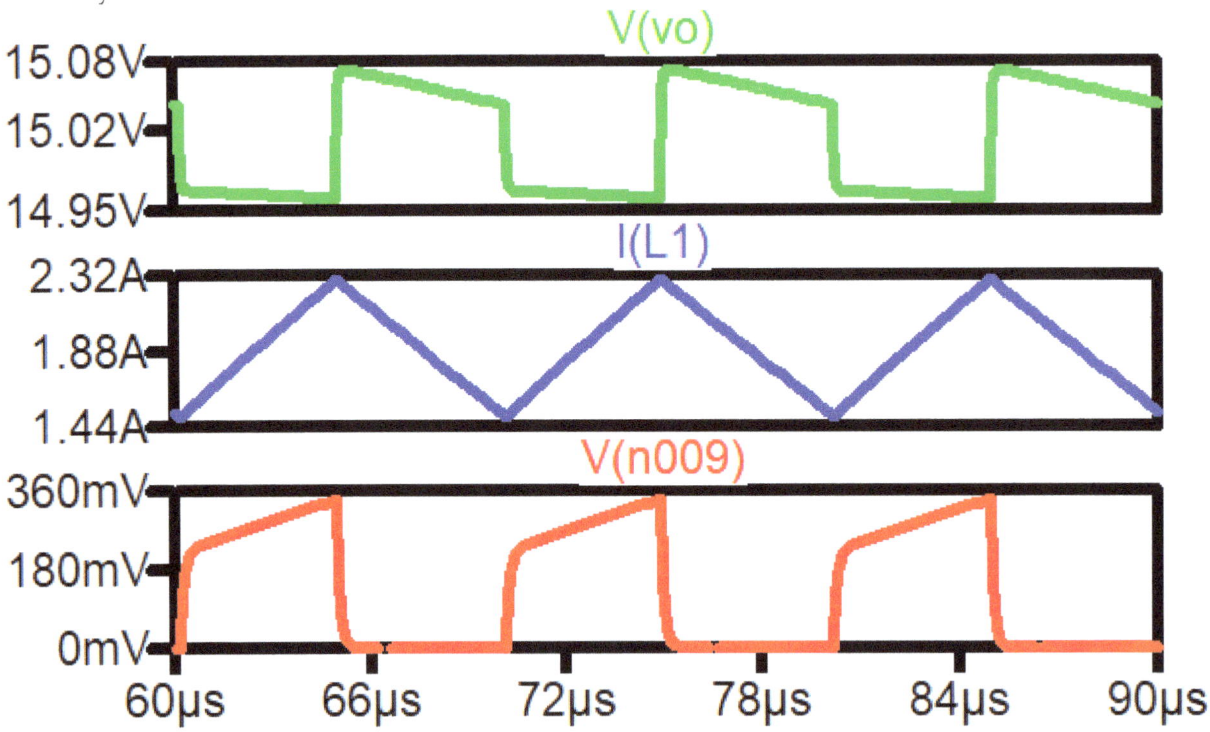

A Cycle-by-Cycle CM Boost Converter

THE CURRENT-MODE version does really differ from the voltage-mode version, which, for safety reasons, also includes a cycle-by-cycle current protection. The main difference lies in the compensator type, which is now a type 2. The crossover frequency does not change, as the upper limit is set by the RHPZ position. However, the transient response with the type 2 compensator, is much better than with the VM version as there is no extra *RC* network across the upper-side resistor.

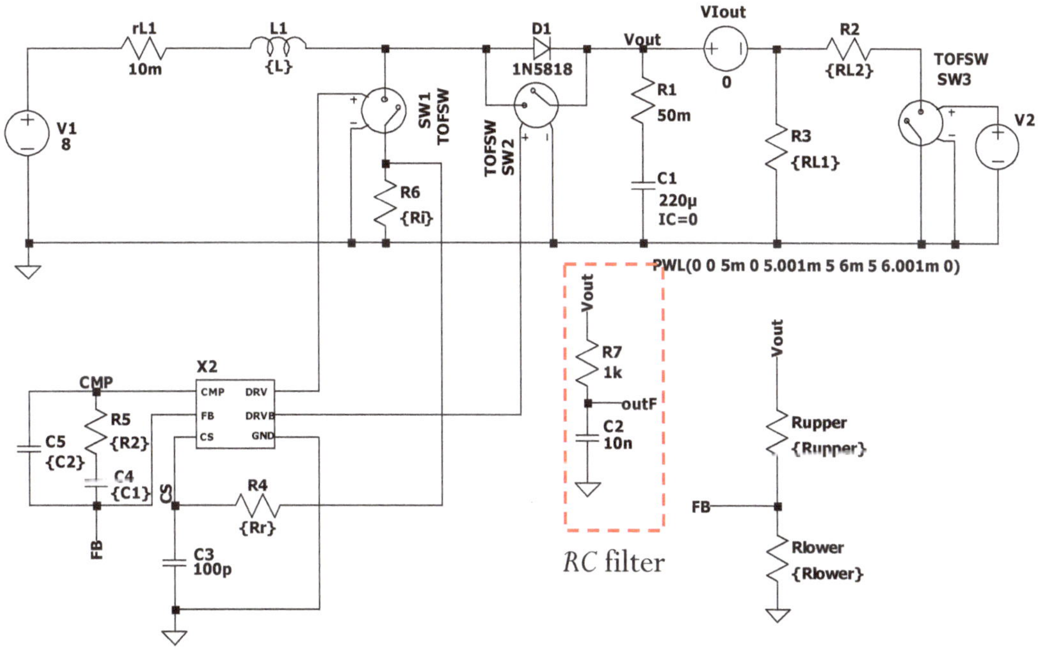

The step load starts from a 70% loading condition and goes up to 100% in 1 µs. The transient response is well-behaved, without overshoot. V_{out} is filtered by an *RC* network for a good-looking graph (too noisy otherwise).

Boost Synchro CM - cbc.asc

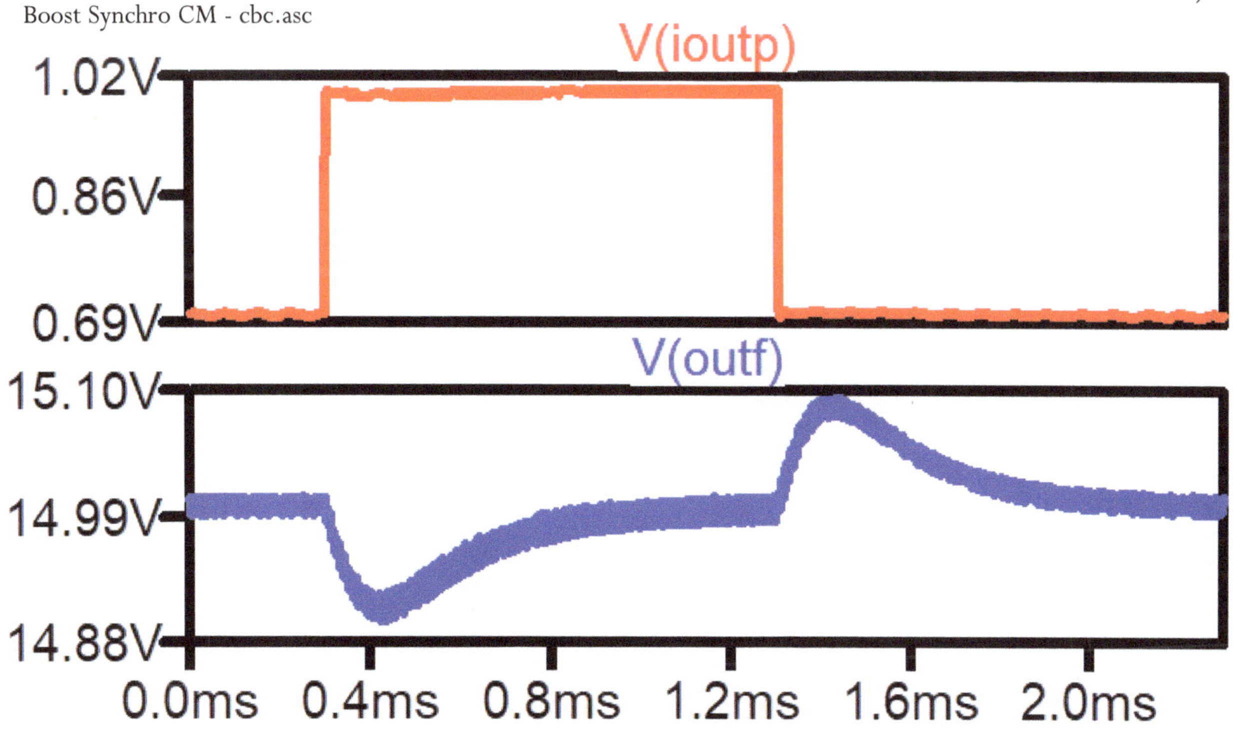

Power Factor Correction in VM BCM

THE BOOST CONVERTER is a popular structure found in power factor correction (PFC) circuits. It is a high-voltage ac-dc converter, purposely controlled to absorb a sinusoidal input current from an ac source. In single-phase applications and for power levels below 300-400 W, the borderline conduction mode (BCM) operation in voltage-mode control brings simplicity and good performance. The designer selects a maximum on-time value proportional to the power being processed. The controller then operates in a self-relaxing mode by detecting the inductor demagnetization end via a dedicated winding. The switching frequency varies with the ac mains period and you can reduce switching losses by synchronizing the turn-on event in the drain valley, like in a QR converter.

The below averaged model corresponds to the NCP1608 from *onsemi*. The error amplifier sits in the right side while the control voltage – actually the on-time value – is computed by B_1.

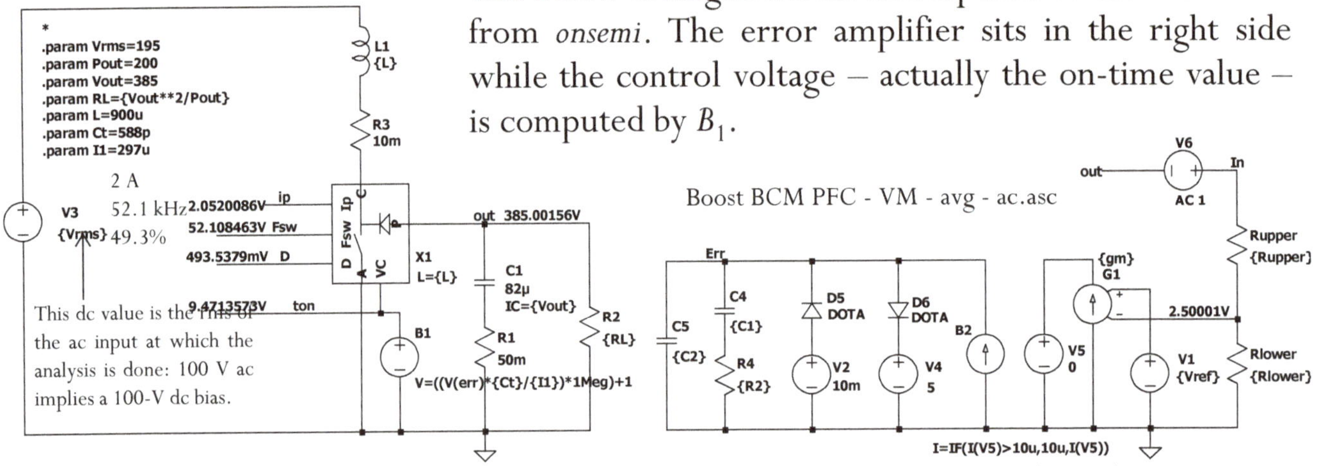

Simulation is fast and shows the various static parameters computed at the operating point: peak current, switching frequency, duty ratio and on-time. You can then run the ac analysis and stabilize the loop.

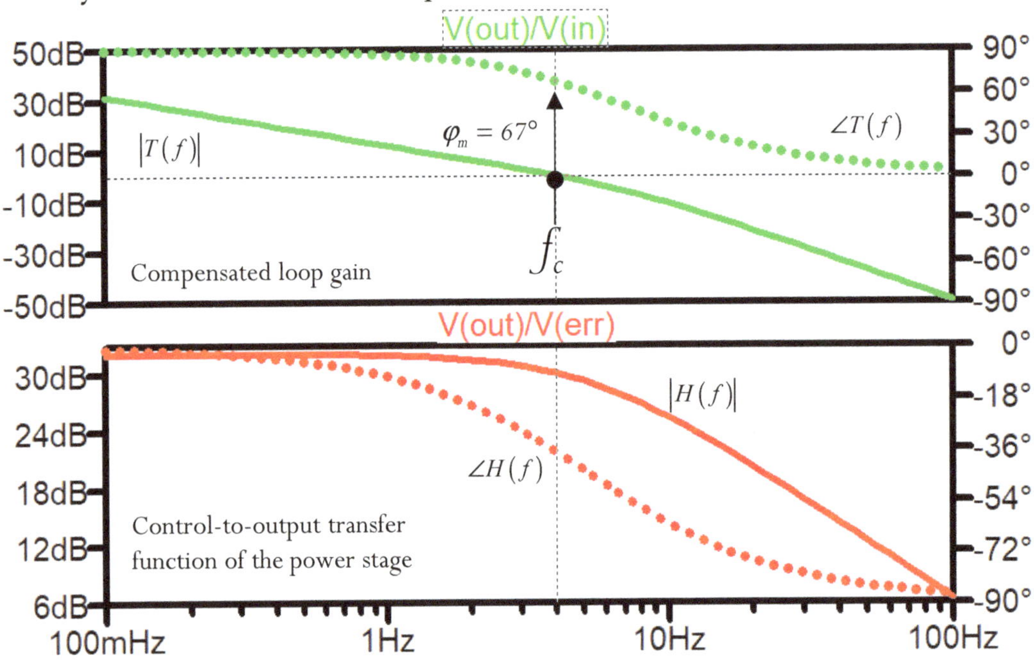

From the power stage transfer function, you feed the macro with the value extracted at the given f_c (4 Hz here, low line) and the loop is automatically stabilized.

Transient Simulations in Averaged Mode

ONE INTERESTING THING with averaged models, is that they can also run in transient mode, owing to their internal large-signal equations. You can then test your compensation strategy and see how it translates in terms of input current harmonic distortion or low-frequency ripple current in the output capacitor. In this mode, the original dc input bias is replaced by a sinusoidal generator.

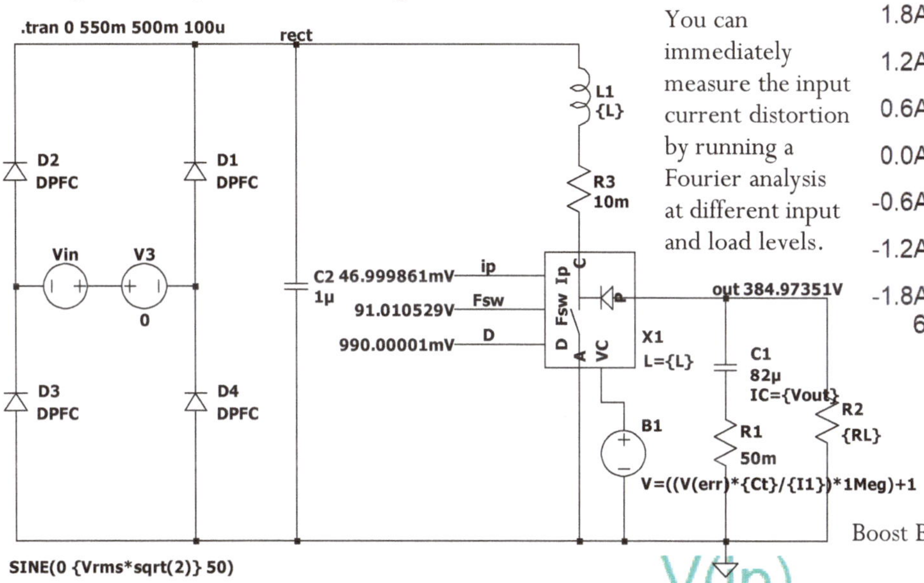

Boost BCM PFC - VM - avg -fixed load.asc

You can immediately measure the input current distortion by running a Fourier analysis at different input and load levels.

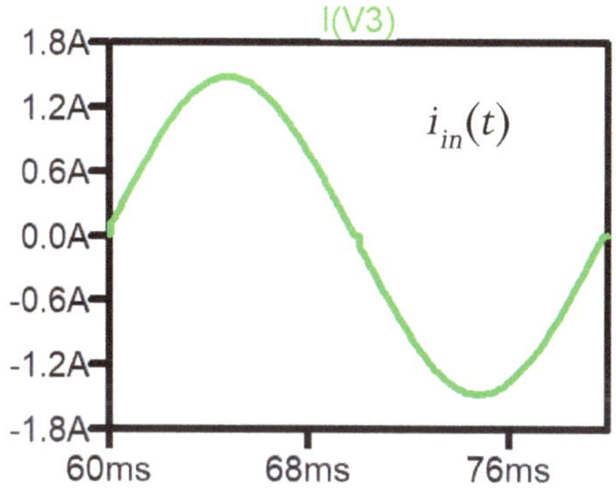

.four 50 10 -1 I(V3)

```
Total Harmonic Distortion:    1.529221%
```

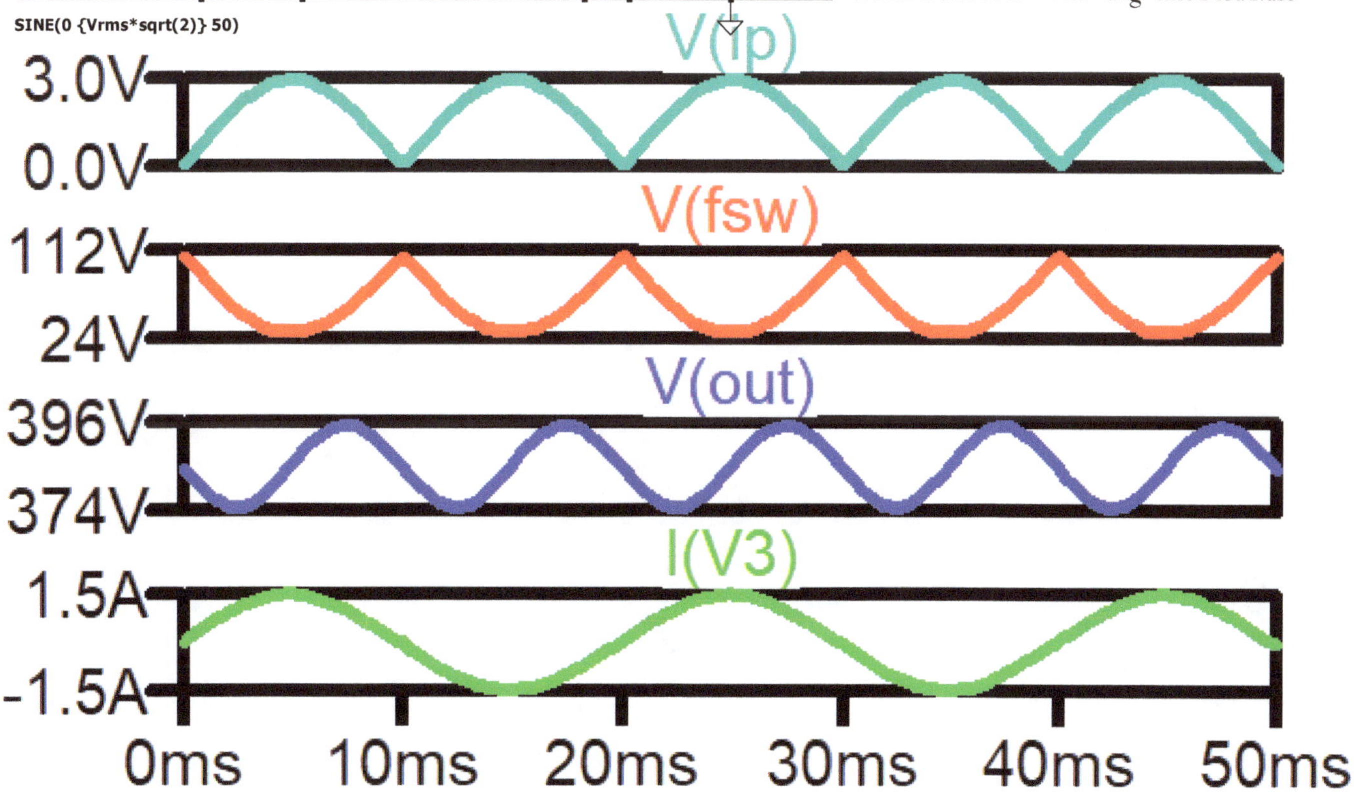

In this graph, you can see the peak current reaching 3 A at low line and the switching frequency varying between 24 and 112 kHz. The peak in frequency occurs at the 0-V crossing event and generates distortion. Techniques exist to improve the THD in this region.

BCM Cycle-by-Cycle Simulations

SIMULATING PFCs in cycle-by-cycle mode, can be a long and tedious process, considering the large time constants involved. The below circuit represents a simplified BCM controller, fast enough, however, to deliver useful results and verify stability in a variety of bias situations.

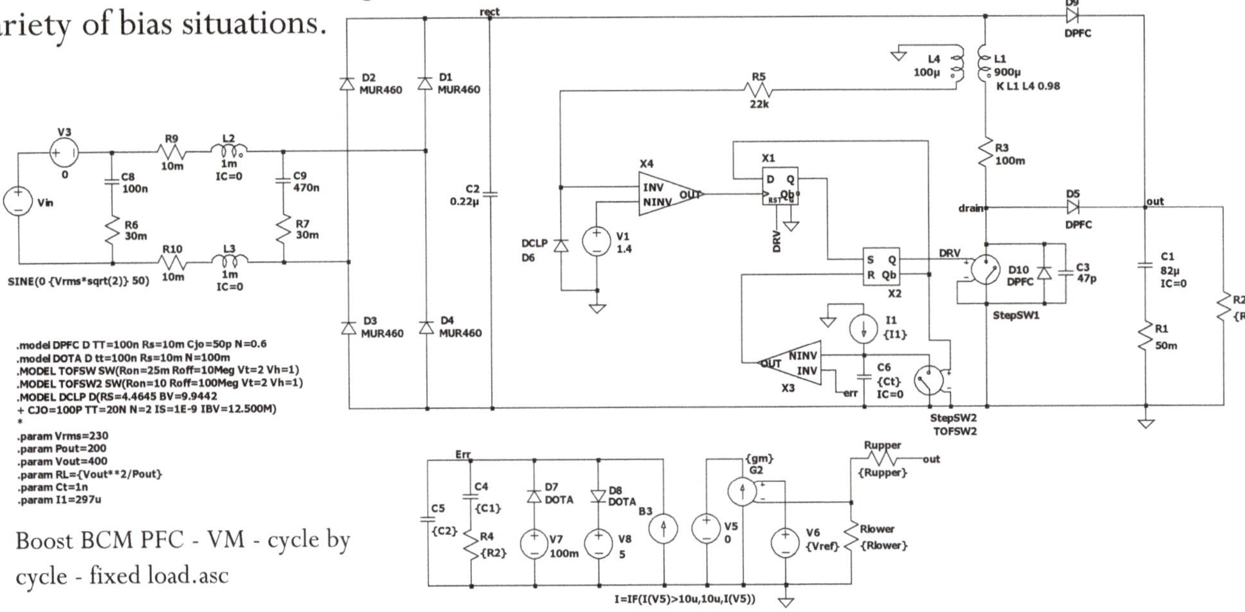

Boost BCM PFC - VM - cycle by cycle - fixed load.asc

Coupled inductor L_4 together with comparator X_4 signal the demagnetization end to restart a new cycle. The on-time generator is built with I_1/C_6 and is reset cycle-by-cycle. The below waveforms confirm the correct operation of the PFC, correctly regulating at 385 V in this example. The simulation time was 100 s, which is acceptable for exploring different operating conditions.

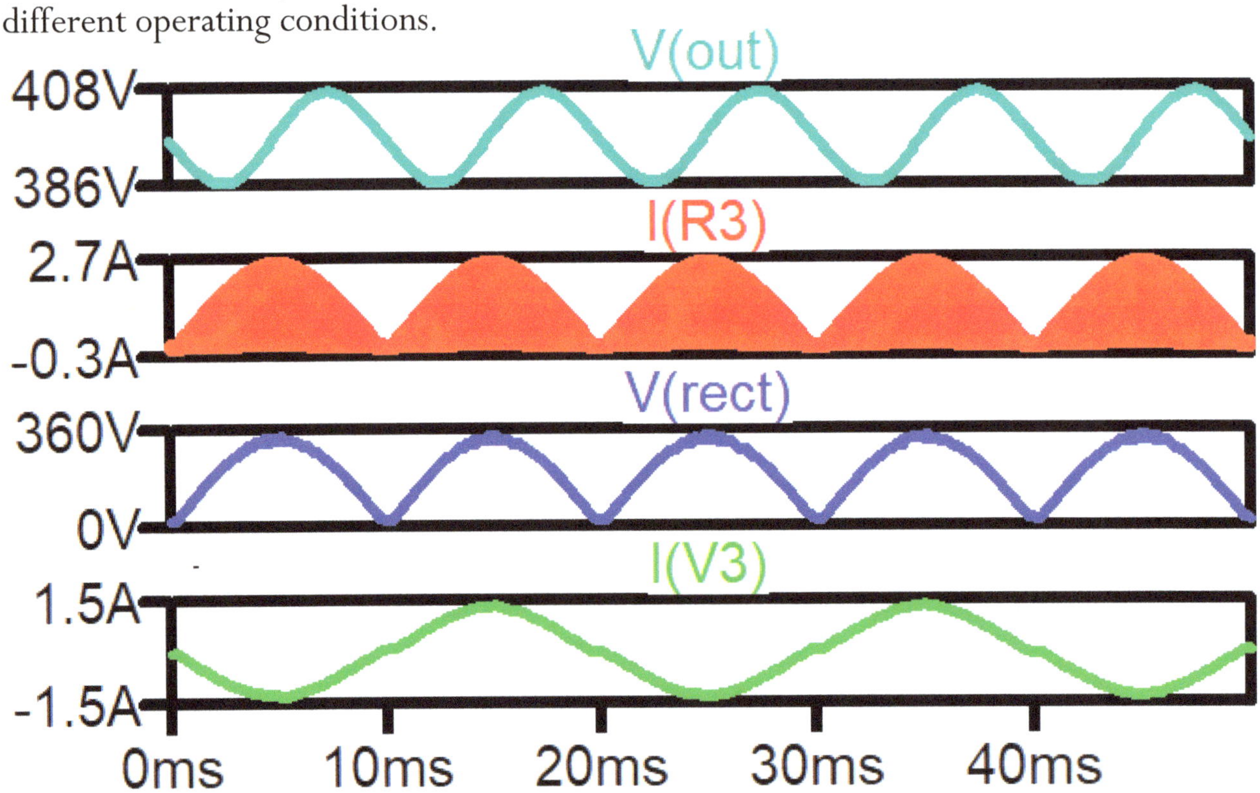

High-Power CCM PFC

FOR POWER LEVELS exceeding 300-400 W, the CCM operation is preferable over BCM: the inductor ripple current is reduced, which induces lower core and conduction losses but, unfortunately, it requires a fast diode with a low recovery charge. Nowadays, a SiC component is well suited for this approach. A CCM PFC can be built in different ways, but the traditional version includes a wide-bandwidth loop forcing the inductor current envelope to follow a sinusoidal shape. The UC1854 from Unitrode was based on this principle and it includes two loops: a fast one for the inductor current, and a slower one for the output voltage regulation. In terms of stability, you must ensure that both voltage and current loops are stable. The below averaged model will help you analyze both sections independently.

In this configuration, the current loop is closed and I ac-sweep the voltage loop via source V_2 in series with the error amplifier. The regulation and compensation are implemented with an OTA, in a type 2 configuration. Its maximum output current is clamped to 28 µA in this example. As usual, carefully check the operating points.

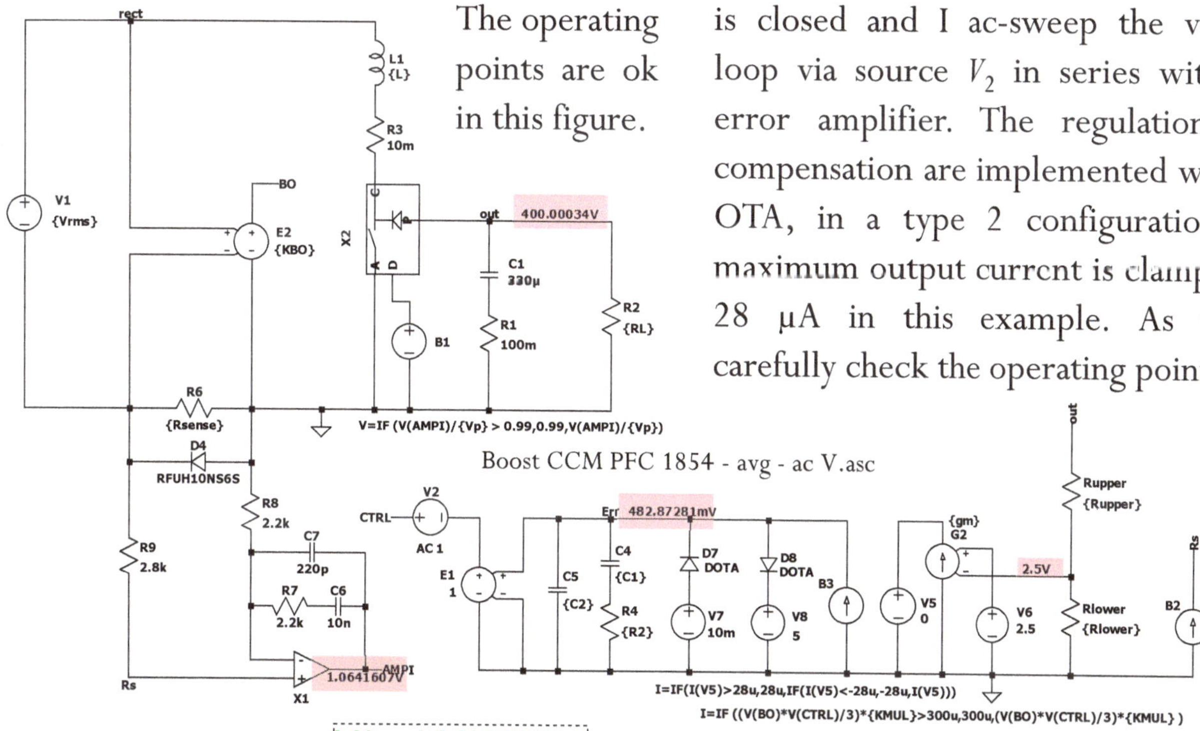

The operating points are ok in this figure.

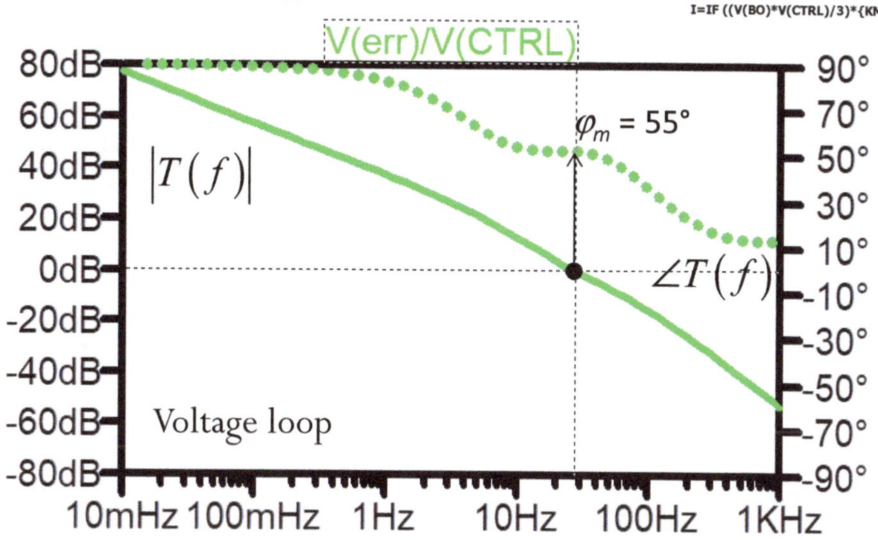

Once compensated, the voltage loop exhibits a crossover frequency of 30 Hz (V_{in} = 230 V rms) and a comfortable phase margin. The UC1854 includes a divide-by-V_{in}^2 that I did not include here. This is to make f_c independent from V_{in}.

Current Loop Analysis

AS EXPLAINED, the internal current-loop must be separately analyzed from the voltage loop. For a low input current distortion, the crossover frequency of this loop must be in the vicinity of 10 kHz, as used in the below example. The ac source V_2 is placed in series with the sensing point, the diode bridge ground. By probing $V(I_{out})/V(I_{in})$, you obtain the compensated gain of the current loop.

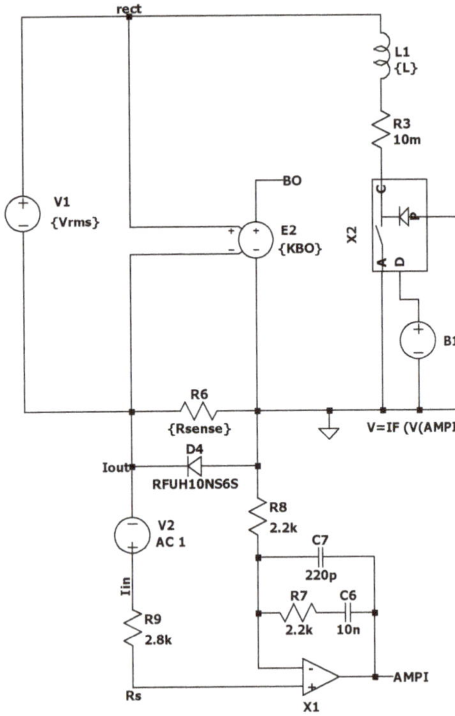

Boost CCM PFC 1854 - avg - ac I.asc

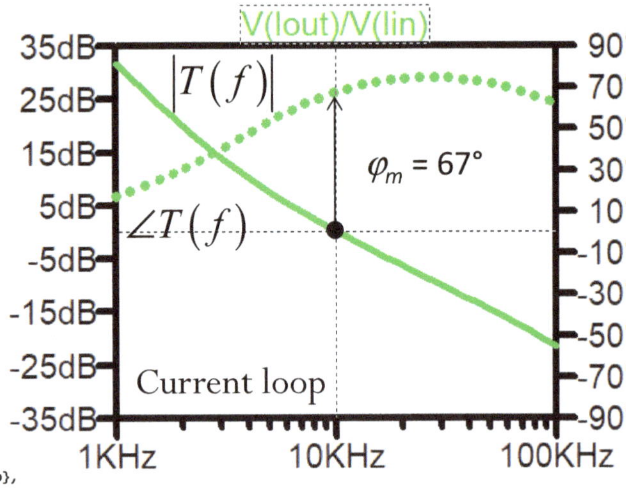

The ac simulation confirms a 10-kHz crossover frequency, and a comfortable phase margin approaching 70°. Now that both loops are stabilized, we can run a transient simulation with the averaged model.

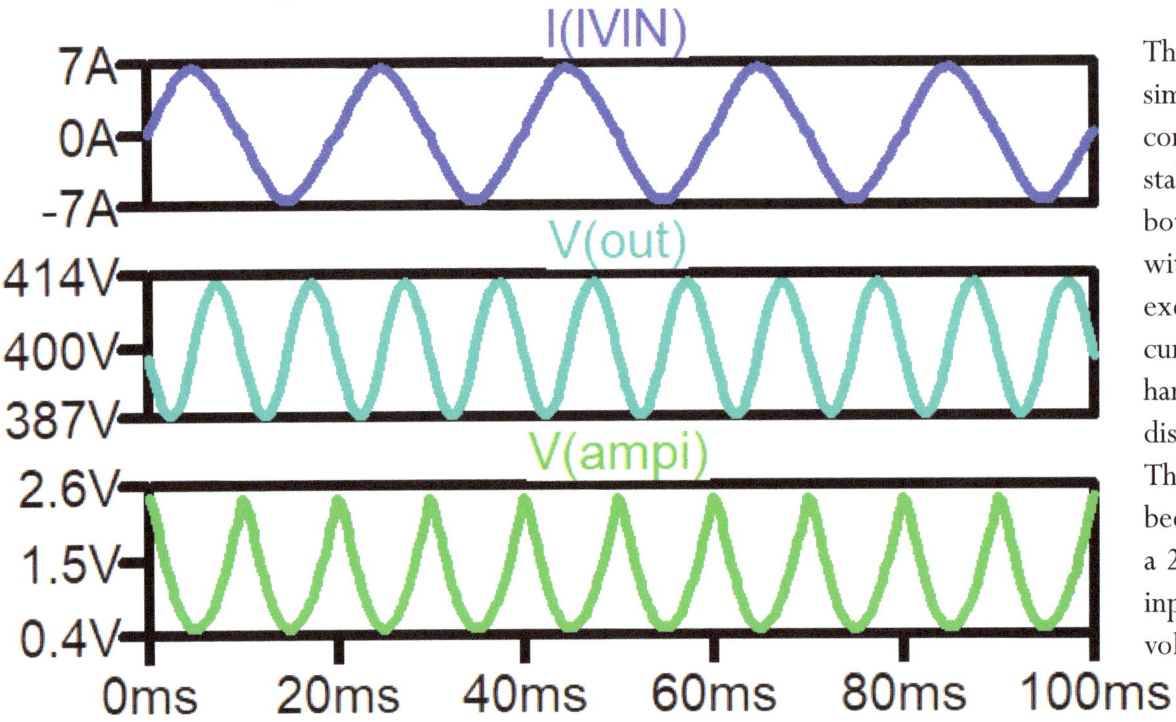

The simulation confirms the stability of both loops with an excellent current harmonic distortion. The test has been run for a 230-V rms input voltage.

Boost CCM PFC 1854 - avg - tran.asc

A Totem-Pole PFC

THE TPPFC structure is known for many years, but the technology was not yet available to make it work properly. Nowadays, GaN and SiC transistors are used in the fast-leg, while slower types, like MOSFETs, rectify the low-frequency mains in the so-called slow leg. I have implemented a simplified TPPFC with off-time control for a high-power CCM operation. The subcircuit includes some decoding logic gates to identify the line polarity and route the control pattern, depending on which of the two fast-leg transistors plays the role of the switch or the diode.

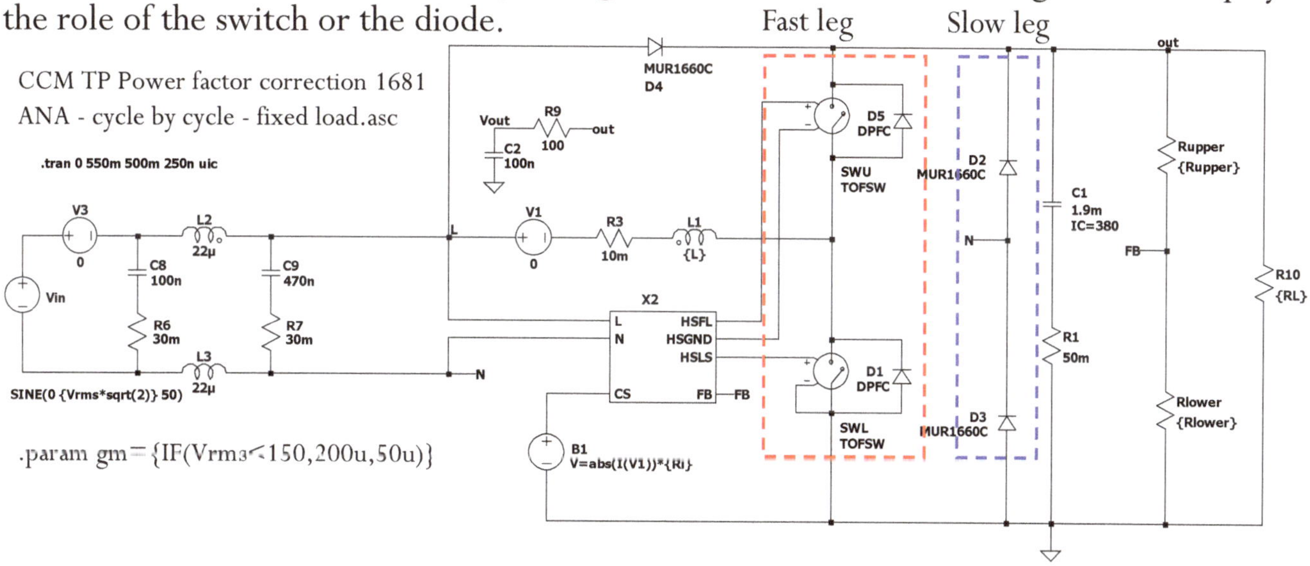

This circuit delivers 3 kW from a 250-V rms input voltage, and its simulation takes a few minutes to complete. The *.param* statement you can see on the schematic, is there to change the OTA transconductance on the fly, depending on the input voltage. This is because the power stage gain will change in low- and high-line conditions, affecting crossover and distortion. Adjusting the type 2 mid-band gain is a way to keep a stable crossover frequency.

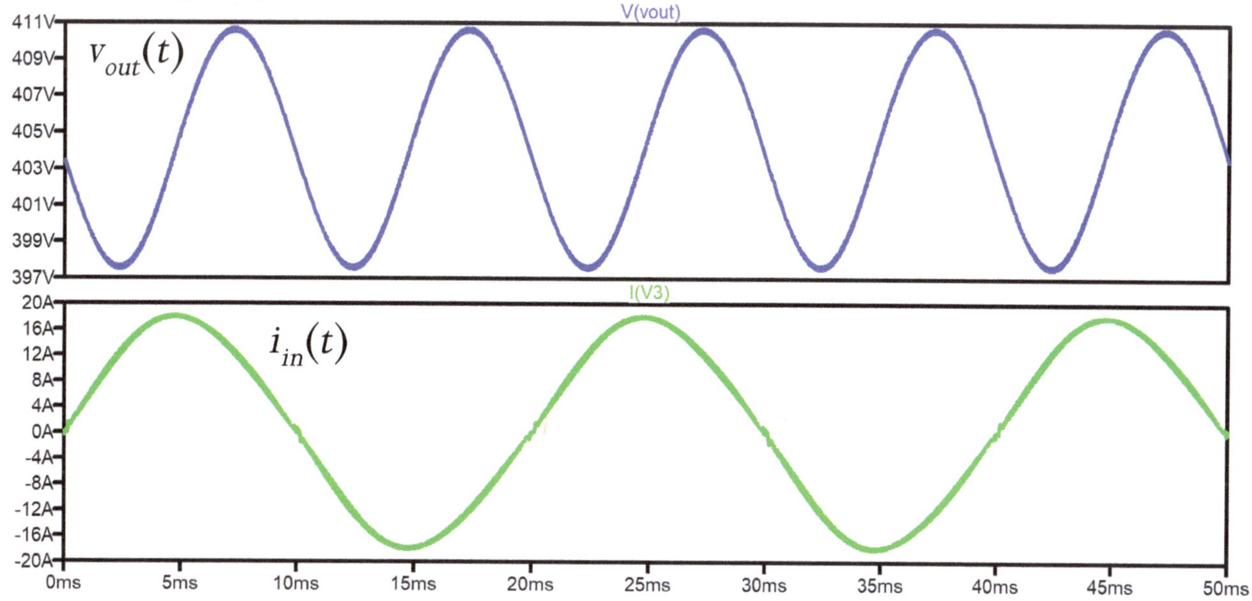

Vienna Rectifier

If we now jump to three-phase PFCs, there are many different circuits, available from moderate to high power levels. The Vienna is one of them and it uses three bidirectional switches in its simplest configuration. Rectifying diodes can later be replaced by synchronous switches if bidirectionality is a wanted feature. Numerous control strategies exist to drive the horizontal switches, and one of the simplest ones, is the variable off-time technique described by Sam Ben-Yaakov in his 1998 paper, *PWM converters with resistive input*. In this fully analog version, input currents are monitored and rectified to feed three wide-bandwidth linear amplifiers.

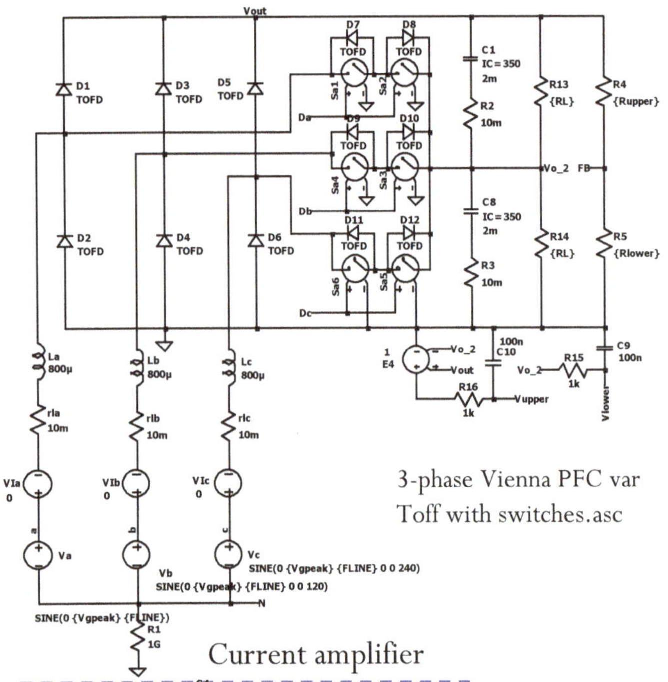

3-phase Vienna PFC var Toff with switches.asc

The Vienna rectifier is a 3-level converter, providing two dc output levels, well suited for stacking downstream dc-dc converters. Considering the two dc levels, the stress on the stacked converters is reduced, and 650-V semiconductors are well suited for a split 800-V rail. Note that the modulation strategy usually includes a secondary loop which ensures voltage balancing between the two capacitors, but I did not include the extra loop in this simple circuit. Here, the converter delivers 10 kW, supplied from a 400-V line-to-line network.

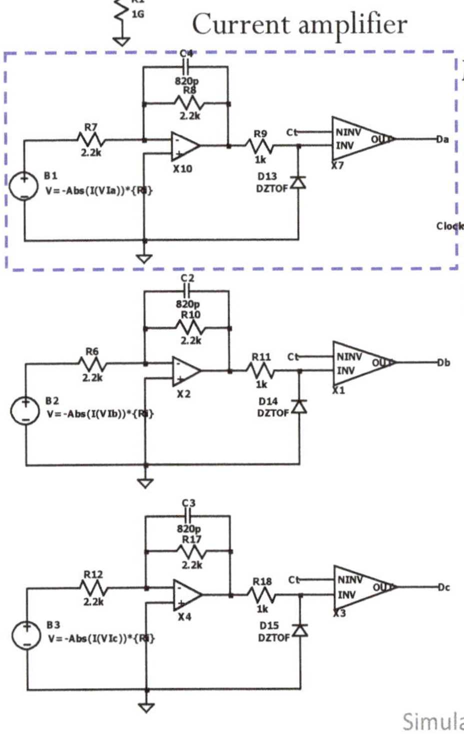

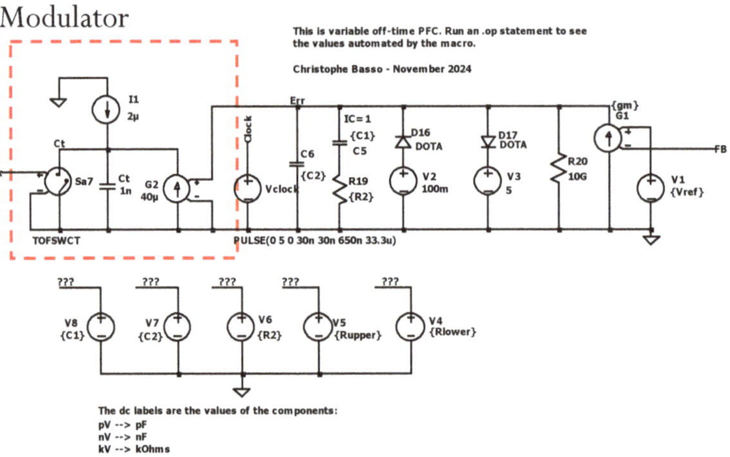

A type 2 compensator built around an OTA, is used to stabilize the converter. The component values can be obtained via the dc sources and a bias point analysis or through a *.options logparams* statement.

3-Phase Vienna Waveforms

The simulation time is fairly quick and you obtain the curves in less than 1 minute, which is fairly convenient. The THD is below 5%, which is very good for such a simple circuit, without any high-voltage sensing, except for output voltage regulation. This is obviously a simplified circuit as most of industrial 3-phase PFCs are digitally-controlled, but I thought it was interesting to present this 100% analog control method.

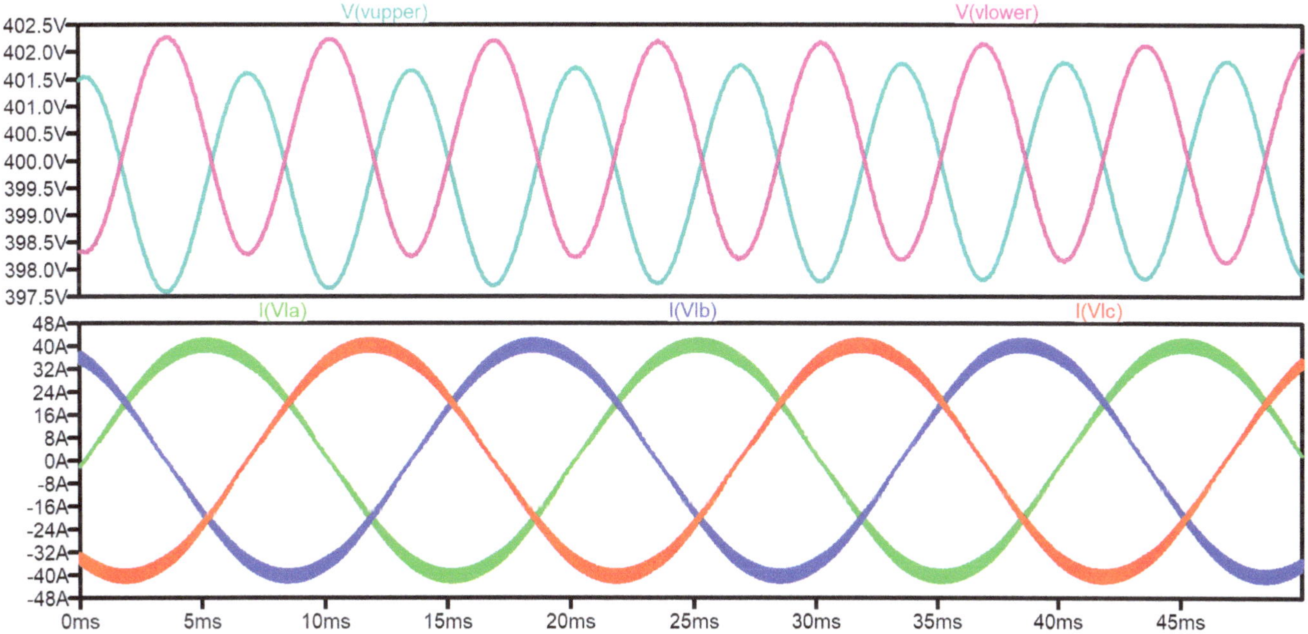

Loop stability can be explored with averaged models as I presented in my webpage. Otherwise, SIMPLIS can simulate the entire circuit and extract the ac response with active alternative sources. A transient load step confirms the whole circuit is stable as you can see below (step from 5 to 10 kW – input voltage = 230 V rms).

https://powersimtof.com/Downloads/PPTs/Vienna_Rectifier_Simulations.pdf

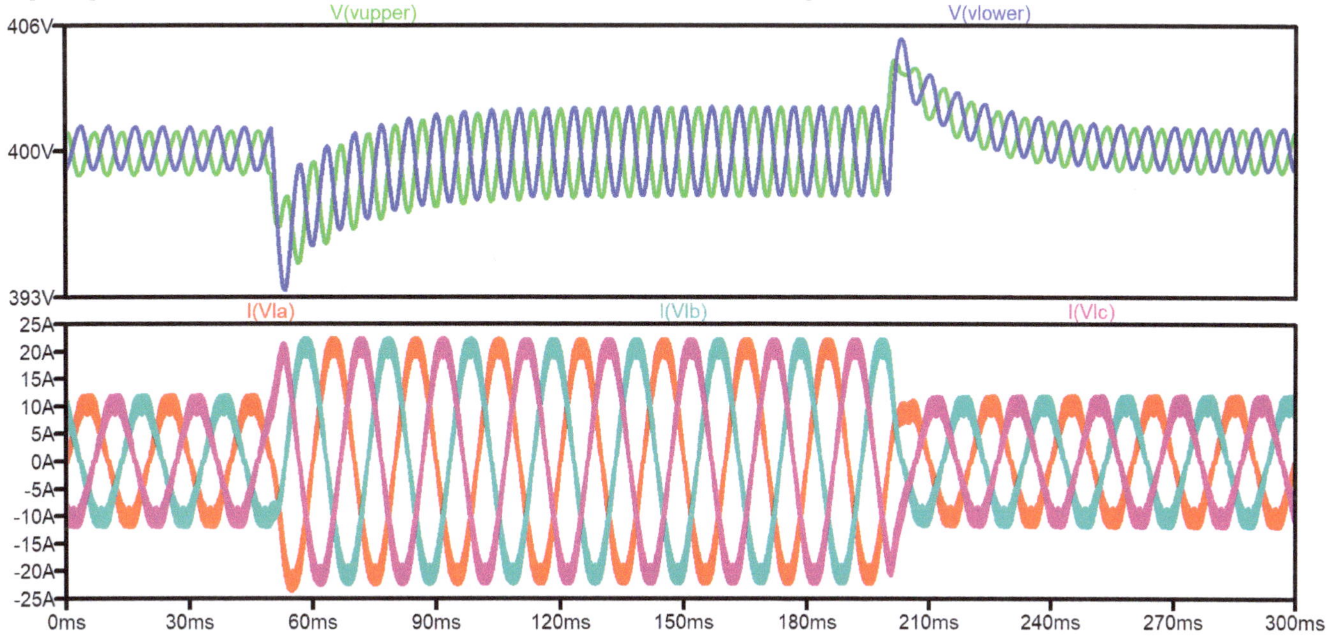

The Buck-Boost Converter – VM

THE BUCK-BOOST CONVERTER is the last one of the three basic switching cells. It offers the worst input/output current signatures, as both are pulsating. The buck-boost converter belongs to the indirect energy-transfer type of structures: during the on-time, there is a first storing step in the inductor, followed by a transfer to the output capacitor and the load at the switch opening. This delay in the conversion process is modeled by a right-half-plane zero (RHPZ), exactly as with the boost converter. You must first calculate its worst-case position – lowest input voltage and highest output current – and then choose a crossover frequency below 20% of this zero.

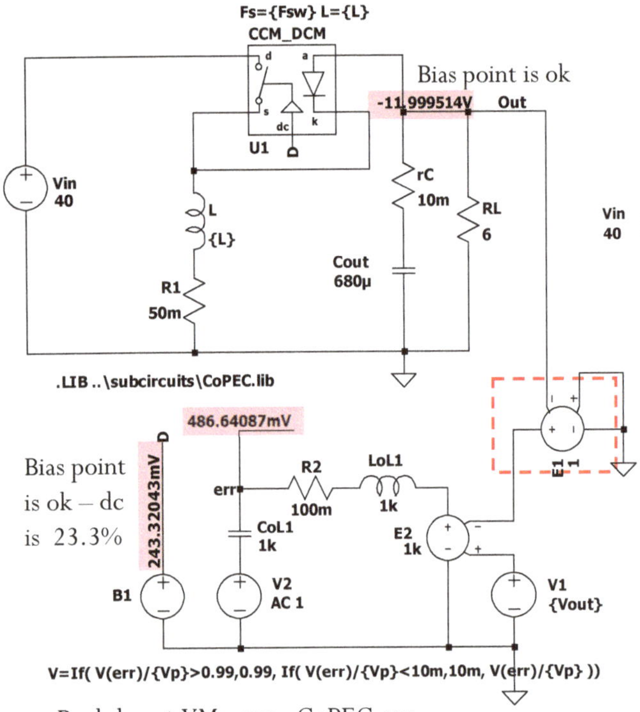

Buck-boost VM - avg - CoPEC.asc

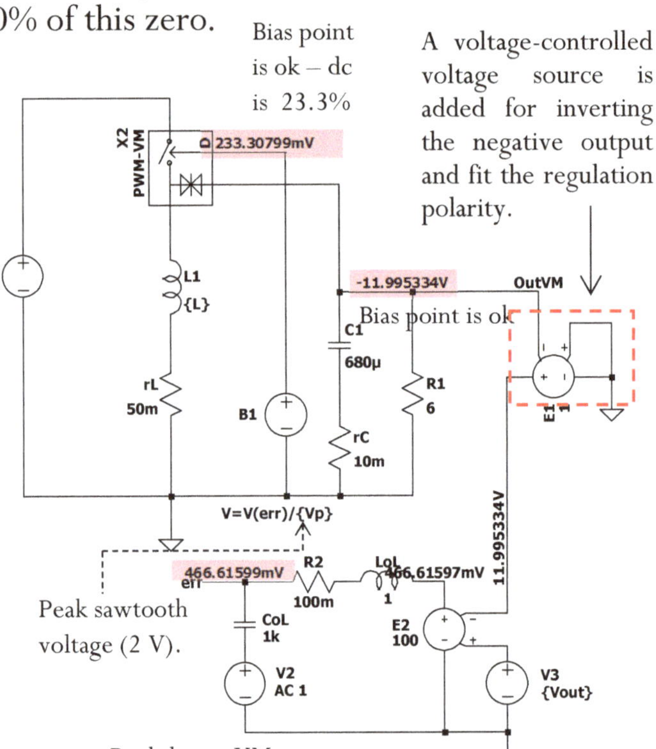

Buck-boost VM - avg.asc

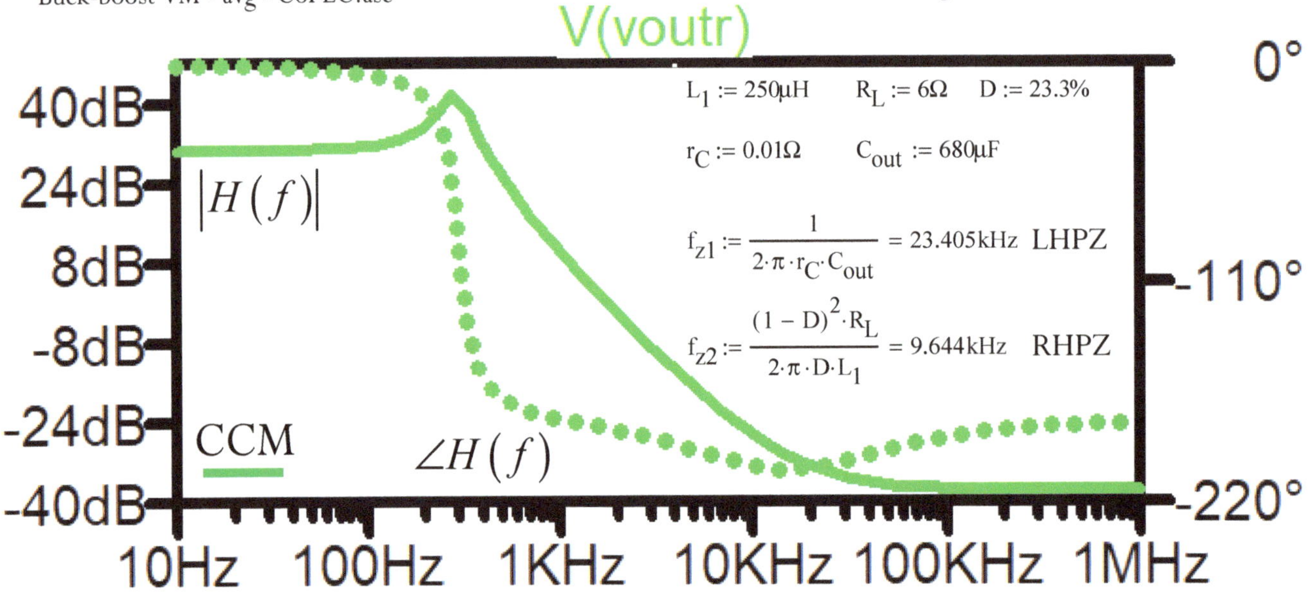

The Buck-Boost Converter – CM

FOR CURRENT-MODE CONTROL analysis, you will also need to reverse the output voltage for proper regulation. Unlike with the boost example, you simply pass a positive value for the sense resistance. You can see that the dc operating point between this CM and the previous VM simulation, are identical. The difference lies in the way the duty ratio is elaborated: *directly* from the error voltage in the VM case, *indirectly* from the peak inductor current in the CM case. This is what changes the dynamic behavior of the two converters.

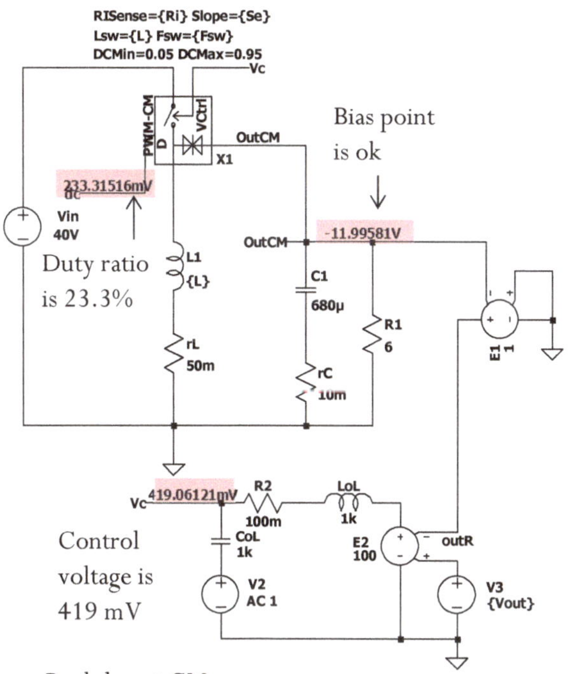

Buck-boost CM – avg.asc

*
.param L=250u ; inductance value
.param Vin=40 ; input voltage
.param Vout=12 ; output voltage
.param Fsw=100k ; switching frequency
.param Ri=0.15 ; sense resistance (neg. for a boost)
.param Sn={Vin*Ri/L} ; on-time inductor current slope
.param Mc=1 ; amount of slope compensation
.param Se={(mc-1)*Sn} ; external ramp for meeting Mc
*

The converter delivers 2 A and there is no slope compensation, purposely to highlight the sub-harmonic poles peaking at $F_{SW}/2$ or 50 kHz. Note that the RHPZ in CM occupies the same position as in VM. However, the *LC* resonance has gone, which gives slightly more freedom to select the crossover frequency.

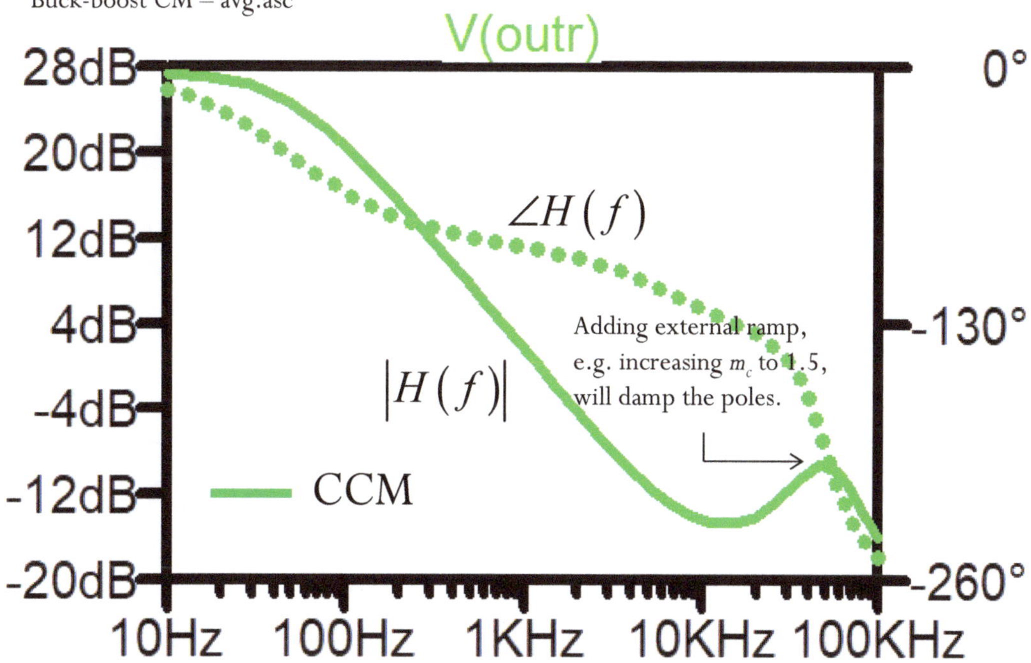

The peaking at $F_{SW}/2$ is lightly pronounced, owing to the moderate duty ratio, below 30%. Pushing m_c to 1.5 will damp these poles even more.

A Simple Buck-Boost VM with FRA

IN ADDITION TO THE AVERAGED MODEL, I have assembled a simple cycle-by-cycle converter featuring a pulse-width modulator. The ramp peak amplitude is 2 V which, together with a 466-mV dc bias, generates a duty ratio of 23.3%. The output voltage is then $0.23 \cdot V_{in}/(1-0.23)$ or ≈12 V if we account for the diode drop. Considering the negative voltage output, I have used the negative port of the Bode box as shown in the schematic diagram. As with the previous open-loop examples, the stimulus amplitude is constant but would be easy to adjust if necessary. I have set it to 30 mV here.

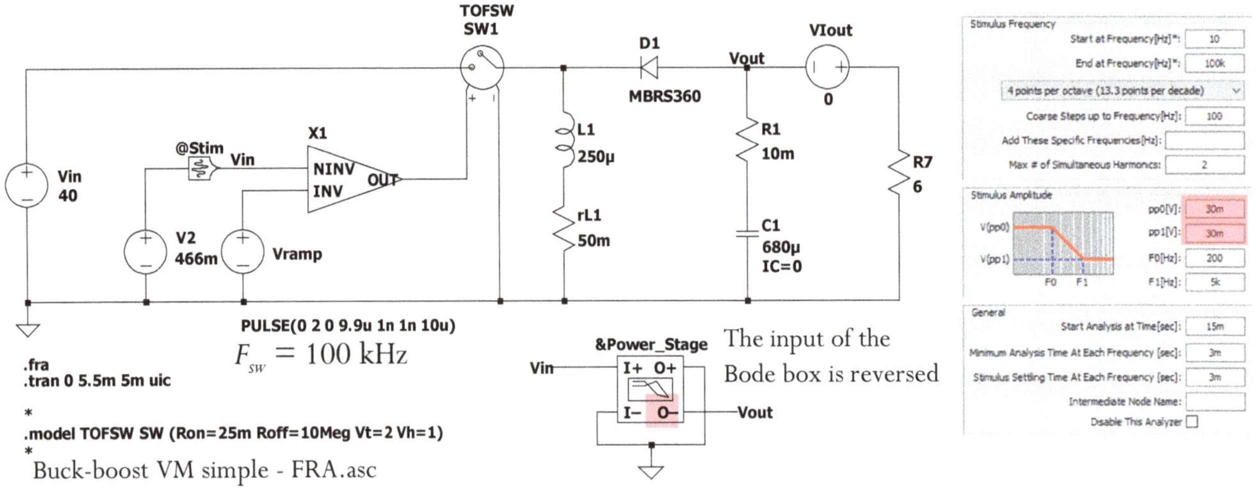

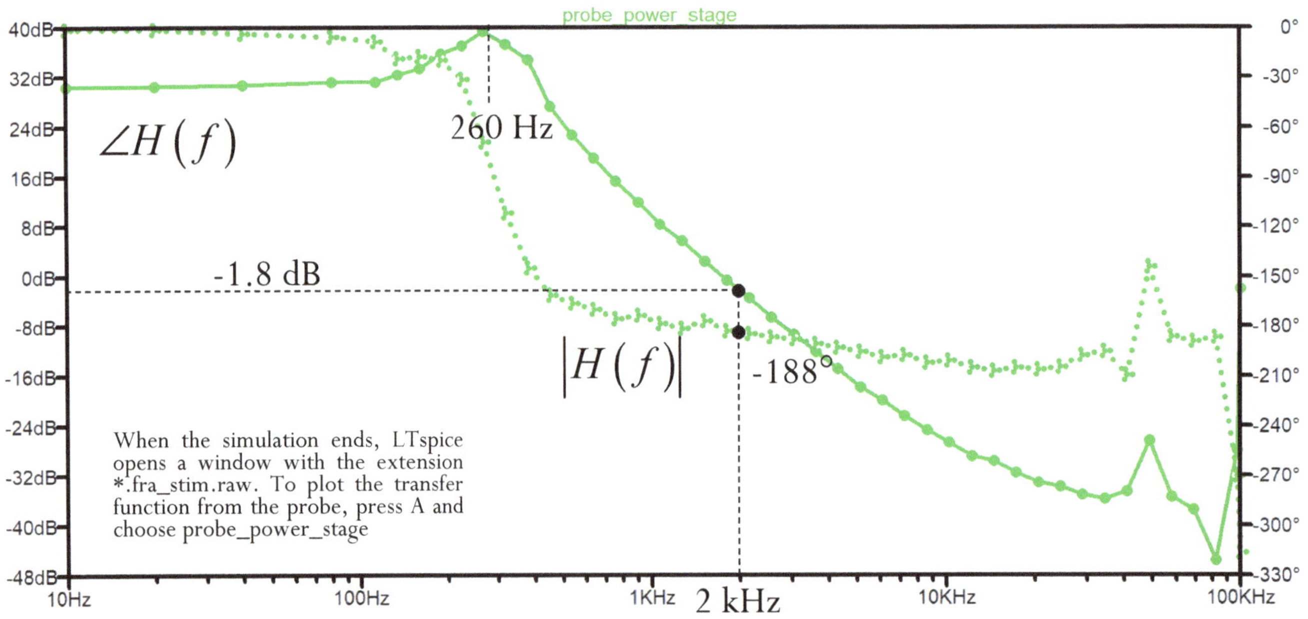

The crossover frequency is selected away from the 260-Hz resonant peak. Considering the RHP zero located at ≈9 kHz, we can select a crossover frequency of 2 kHz. We will have sufficient gain at 260 Hz (the peak will move right at higher input voltages) and we are still below the RHP zero lowest position (minimum V_{in}, maximum output current). Averaged and cycle-by-cycle simulations will confirm whether this is a good choice or not. A better option would be to use to CM control and with it, forget the peak at 260 Hz.

Compensating the Buck-Boost Converter

WE CAN TAKE ADVANTAGE of the averaged model to do a compensation exercise on the voltage-mode-controlled buck-boost converter. We have plotted the control-to-output transfer function and we can now think of a compensation strategy. In a practical application, voltage inversion is internally done to allow negative-voltage regulation. In this example, I have used the same configuration as with the boost converter and all calculations are automated. In this manual approach – where you place the two zeros first – it is important to check the components values in the LOG file. If one of them is negative, it means the zeros are not low enough and the macro cannot meet the phase margin target: lower the zeros position or be less ambitious in terms of phase margin. Here, I have selected a 2-kHz crossover with a 60° PM.

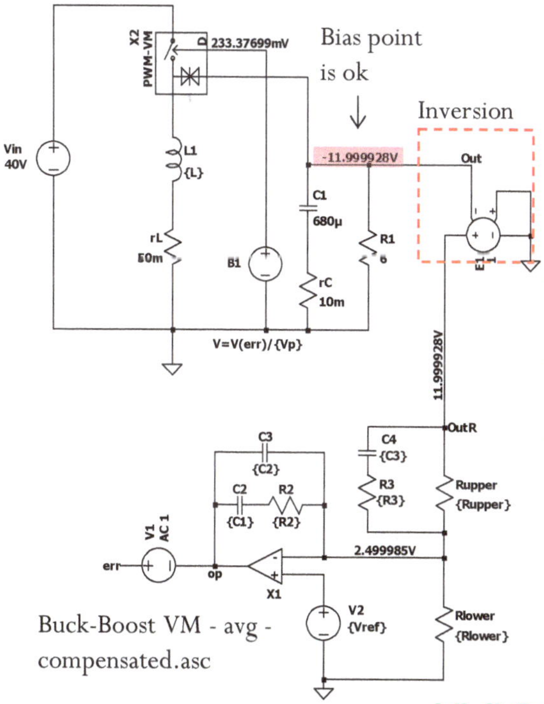

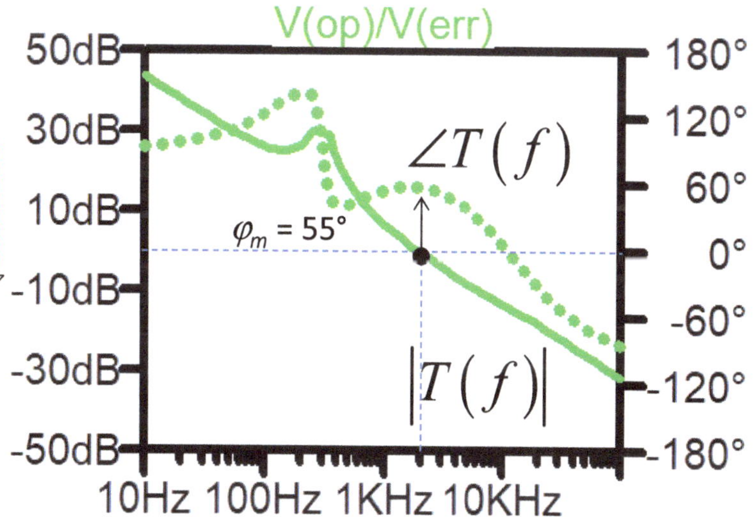

The compensated loop confirms the 2-kHz crossover frequency with a decent phase margin of 54°. Let's have a look at the transient response.

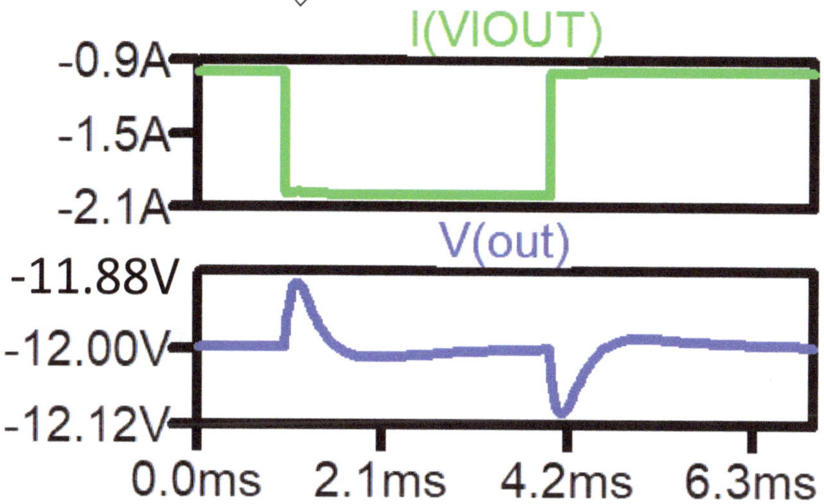

The output current is stepped from 1 A up to 2 A, leading to a drop below 150 mV. In this example, the inductance could be decreased, which would shift the RHP zero higher, giving the opportunity to push the crossover frequency a bit farther.

A Switching VM Buck-Boost Converter

THIS EXAMPLE implements the converter supplied from a 40-V rail, with an output current is 2 A. The sensing circuit with B-source B_1 checks that the switch current remains below a value set by R_i, cycle-by-cycle. The maximum sense voltage I_{max} of 1 V implies a peak current of 1 V/250 mΩ = 4 A.

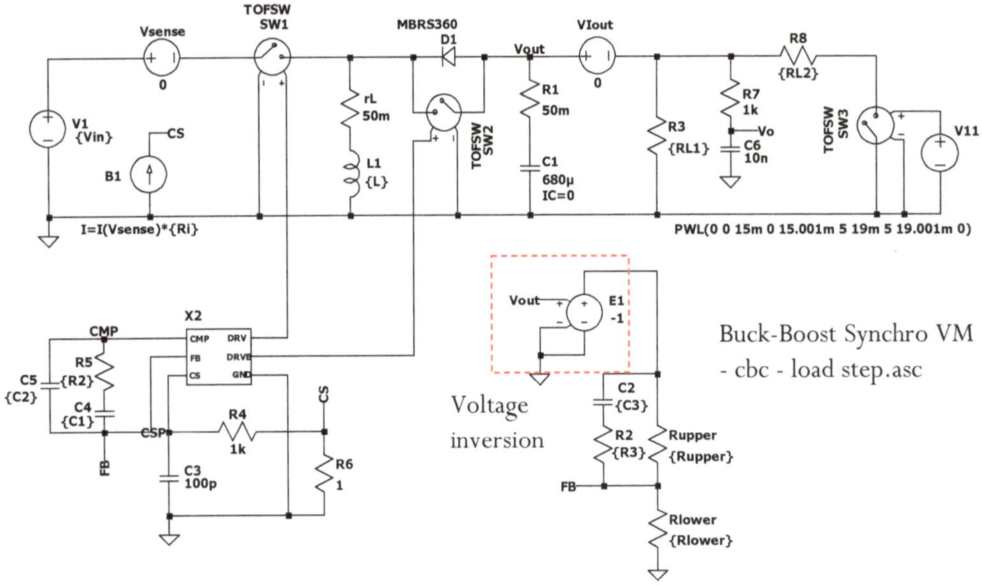

This type of simulation helps you check important operating parameters such as inductor ripple current, input/output capacitor rms current etc. I have added a small R_7C_6 filter so that switching noise does not pollute the observation of V_{out}. I used a controlled switch to toggle the loads, but a PWL current source would also do the job well.

As shown below, the transient response is stable and confirms the study we did with the averaged model. As a recommendation, it is good to first stabilize the loop with the averaged model for the ac response but also for the simulation speed. Once stability is ensured, confirm with a transient load step that the response is well behaved and within expected boundaries.

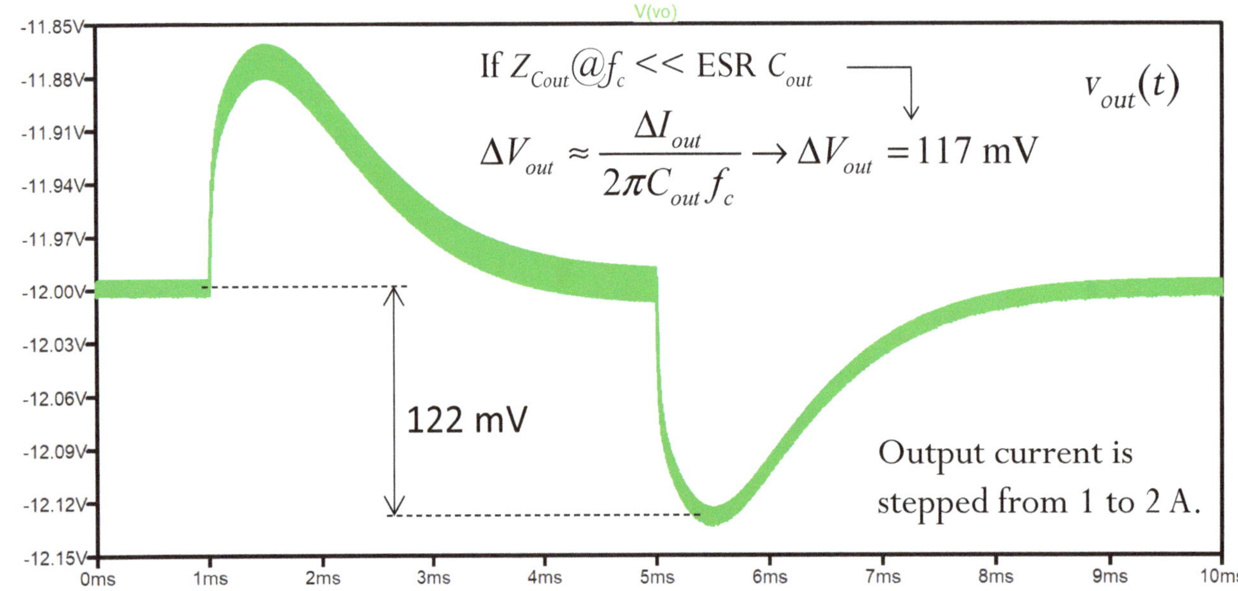

If the output capacitor ESR is way smaller than the capacitor output impedance at crossover (2 kHz), we can approximate the output undershoot during the 1-A step [6].

A Switching CM Buck-Boost Converter

IN CURRENT-MODE CONTROL, the control-to-output transfer function of the buck-boost converter becomes that of a 3rd-order converter. A pole is located at low frequency and two sub-harmonic poles take place at $F_{sw}/2$. The RHP zero occupies the same position as in voltage-mode control, but the resonant peak linked to the *LC* network no longer exists. The compensation is done with a simple type 2 compensator while, depending on the operating duty ratio value, the sub-harmonic poles may require damping.

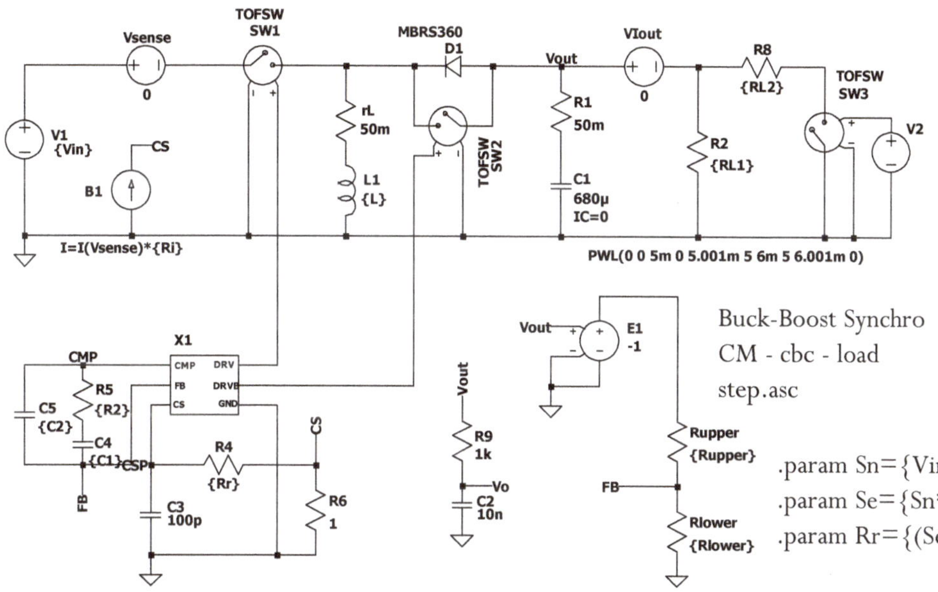

Slope compensation is obtained by inserting resistance R_4 in series with the current-sense information. Combined together with the internal 19-kΩ pull-up resistance, the macro computes the resistance value R_r for a quality factor of 1. In case no compensation is required, R_4 is set to 100 Ω by default.

.param Sn={Vin*Ri/L} ; current slope
.param Se={Sn*(0.818/(1-D)-1)} ; Se for Q of 1
.param Rr={(Se/Sramp)*19k+100}

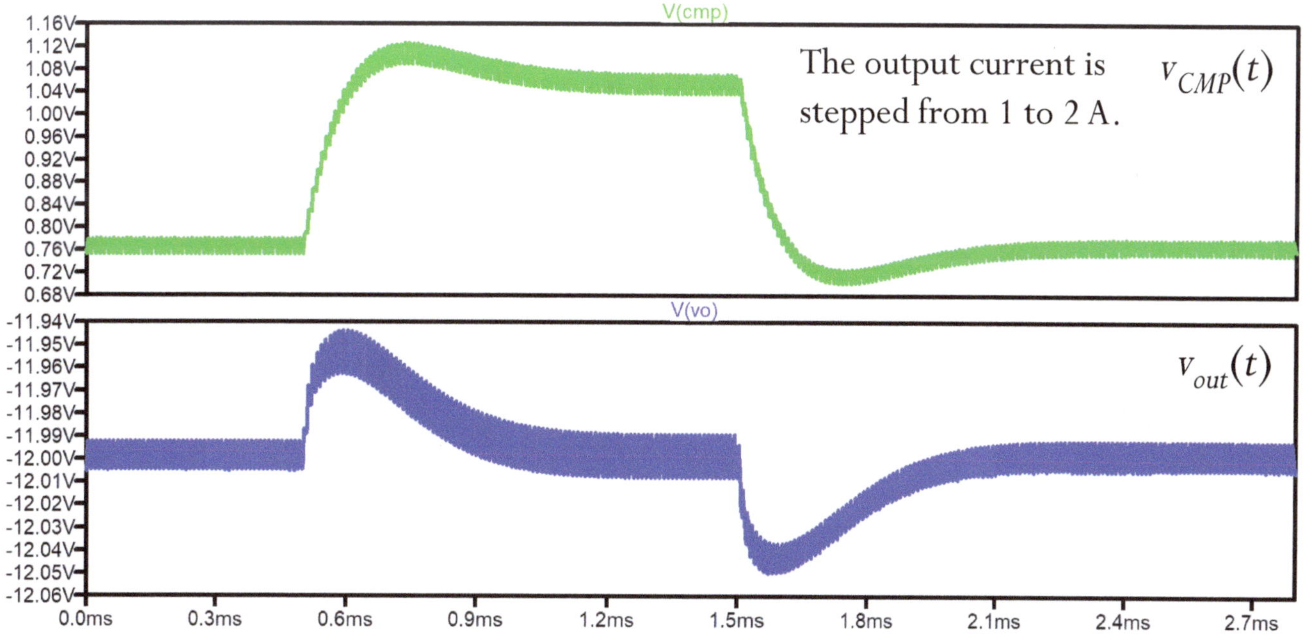

The transient response is excellent, with a smaller drop than with the VM converter. In absence of the *RC* network across R_{upper}, a type 2 compensator usually behaves better than its type 3 counterpart when the converter is subjected to a load step.

Current-Mode Flyback Converter

THE FLYBACK CONVERTER is found in a variety of applications, ranging from a few watts up to 150-200 W. Operated with a single switch in the simplest approach, it can be controlled in voltage- or current-mode, and operate at a fixed or variable frequency. We will start with a simple CM converter as VM is not a popular option. Before going to a cycle-by-cycle simulation, I recommend that you confirm the operating point is correct then explore stability with an averaged model. You will immediately see if the calculated component values (turns ratio, peak current etc.) fit your power requirements down to the lowest input voltage. Below is the typical version of an isolated ac-dc converter found in the consumer market. The feedback is made by combining a TL431 with an optocoupler in the classical way. Regarding the latter, you need to characterize it and extract its pole and the parasitic collector-emitter capacitance which affects the pole position in a type 2 compensator.

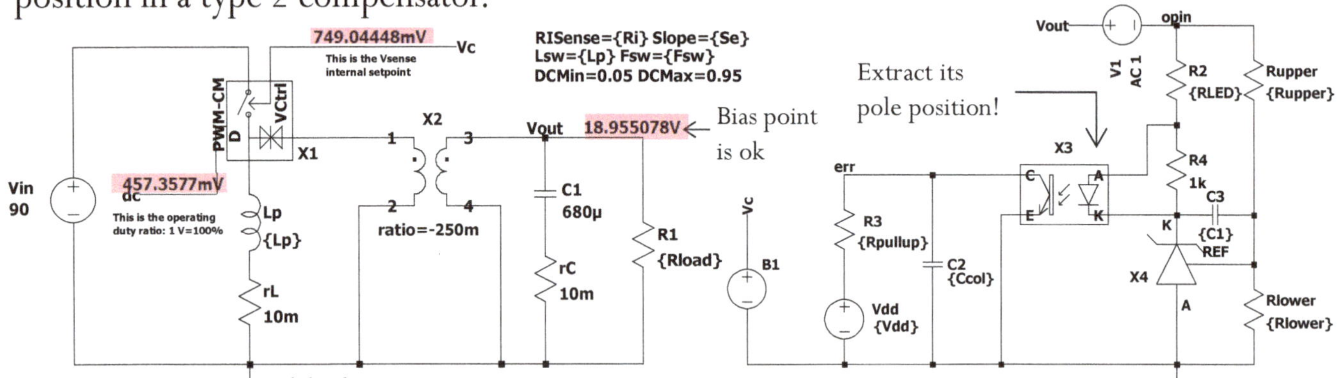

Flyback CM - avg - opto TL431.asc

Before selecting the crossover frequency, you have to determine the lowest position of the RHP zero. If you deal with an ac-dc converter, the rectified valley ripple is the lowest V_{in} (with margin) at which you calculate the RHPZ. Otherwise, it is the minimum dc input voltage with a classical dc-dc converter. Check the LED current on the TL431 – R_{max} value in the macro – to ensure a proper bias with low output voltages (5 V and below). If you don't have margin, then remove the fast lane for an easier implementation.

```
.param D={Vout/(Vout+N*Vin)} ; duty ratio calculation
.param mc={0.818/(1-D)} ; recommended compensation value for a Q of 1
.param Sn={(Vin/Lp)*Ri}
.param Sramp={2.5/Tsw} ; 2.5 V over Ts - check your IC specs
.param Se={(mc-1)*Sn}
.param fRHPZ={((1-D)**2*Rload/(D*Lp*N**2))/(2*pi)}
.param fcMAX={0.2*fRHPZ} ; maximum crossover frequency
```

Check values in the LOG file with the *.options logparams* statement.

The macro will determine the necessary amount of compensation ramp to properly damp the sub-harmonic poles. The RHPZ position is then computed and I recommend a maximum crossover frequency f_c below 20% of the zero position. As underlined, make sure the valley voltage of the bulk capacitor ripple is selected at the minimum V_{in}.

Flyback Converter – AC Response

ONCE THE OPERATING POINT is confirmed to be ok – I designed this converter for a 19-V output voltage – you can move on to the ac analysis. By probing different nodes, you can display the power stage *H*, the compensator *G*, and finally, the loop gain *T*. From the power stage response, you extract the magnitude and phase data at the selected f_c then feed the macro. You can now check the type 2 compensator response and confirm that the loop gain crossover frequency and margins are all acceptable. Deviations naturally exist between what you want and what is measured. For instance, the TL431 offers a low open-loop gain compared to an op-amp. The dynamic resistance r_d of the optocoupler LED together with the extra 1-kΩ resistance (for biasing purposes) also slightly reduce the original gain, depending on the operating point.

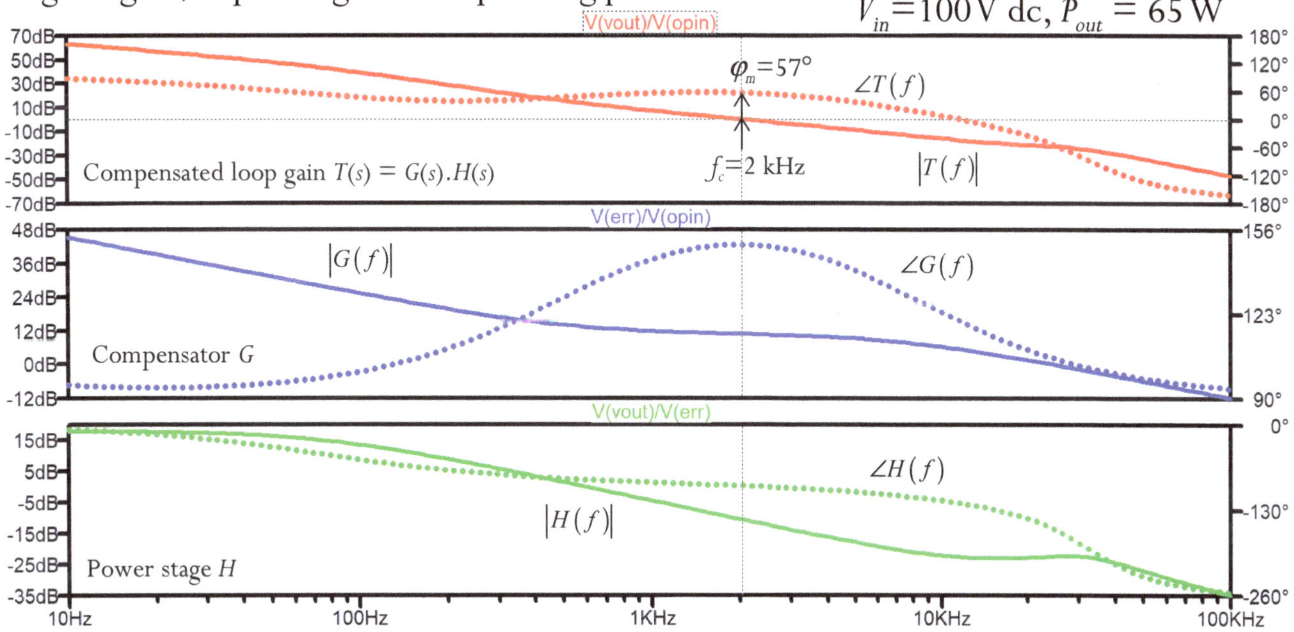

The transient response is excellent and confirms stability is good in this example.

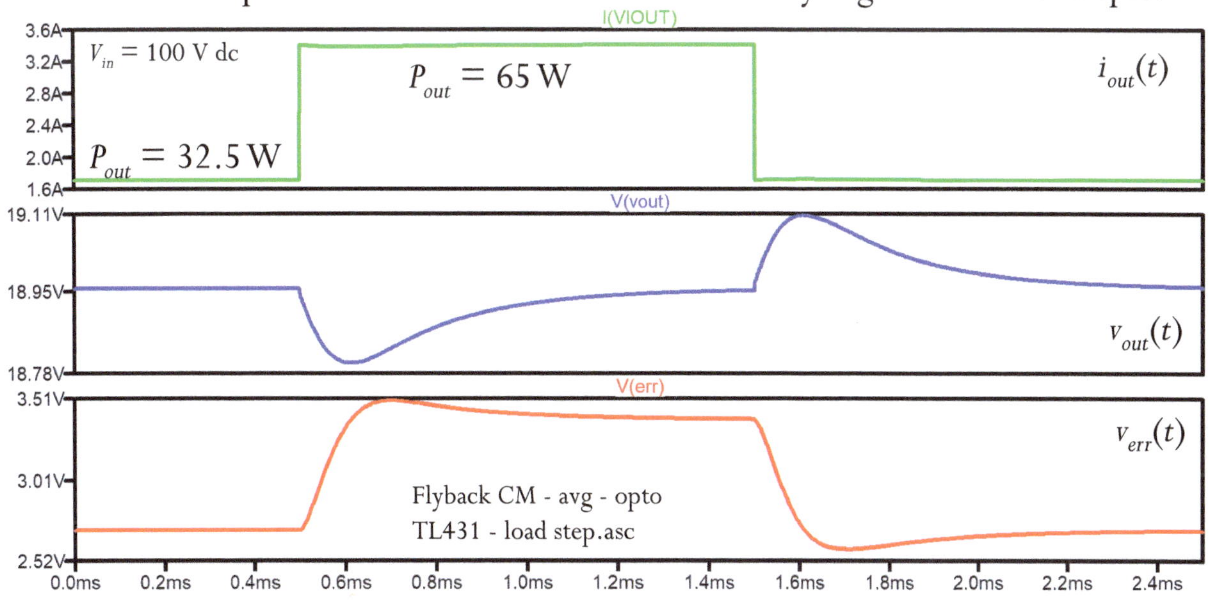

Flyback Converter – AC Input

NOW THAT WE HAVE our loop closed, we can check the cycle-by-cycle version. I have captured many different versions for you to play with. I will show the one built around the UC3842 which features an auxiliary winding, a start-up resistance and an ac input. Slope compensation is externally provided through the DRV pin and the *RCD* network.

Flyback UC384x - aux W MOSFET - isolated - cbc AC in.asc

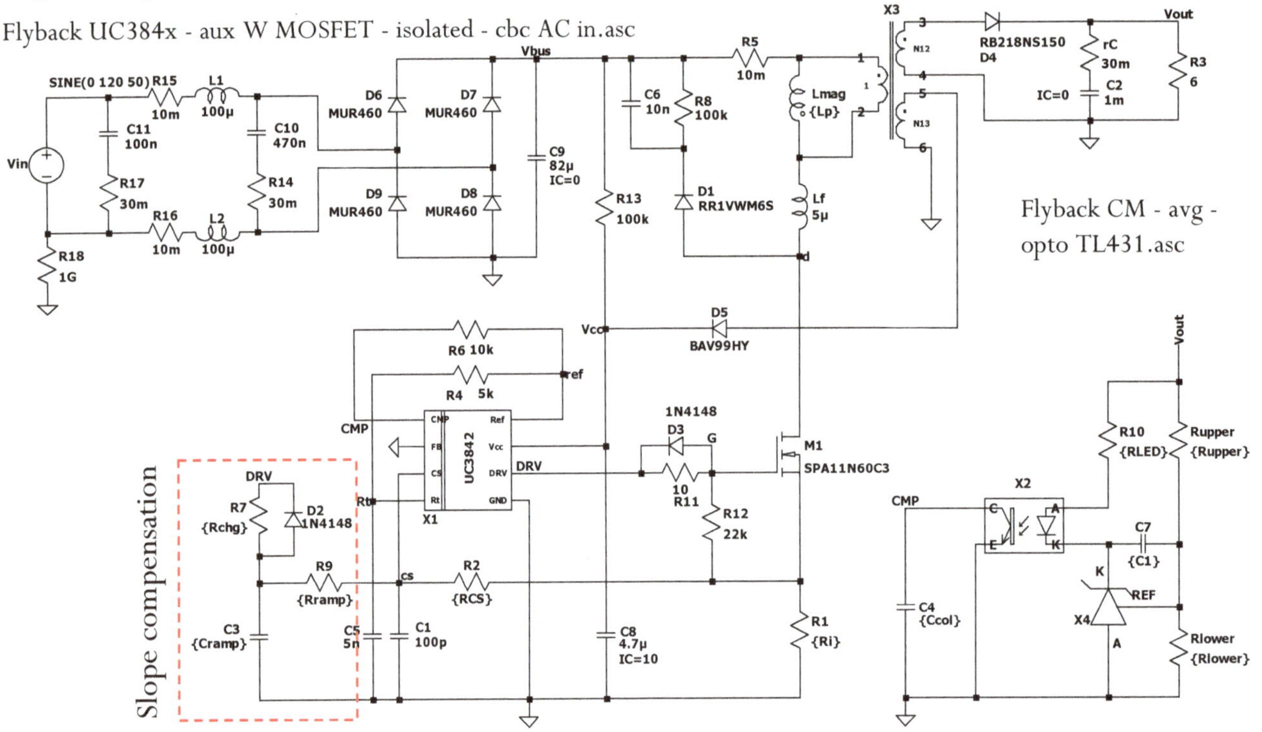

The simulation is fairly fast despite the 50-Hz input sinewave. The input voltage is 85 V rms and you can see the low valley voltage of 60 V in the dc bus. You realize that the power converter must deliver 100% of the nominal power – and be stable – at this low input voltage. The below waveforms show a bit of ripple on V_{out}, but still acceptable.

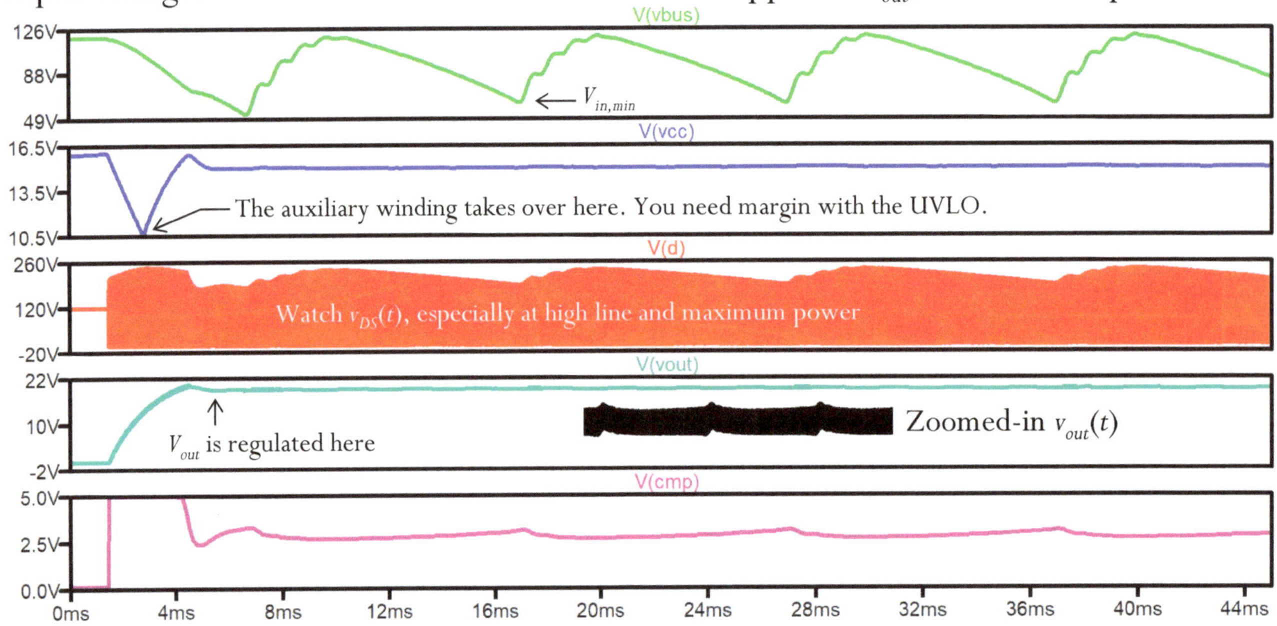

Flyback Converter – Frequency Foldback

MODERN PWM CONTROLLERS include a reduction in the switching frequency when the load is becoming lighter. This action reduces switching losses and improves efficiency when the power supply is less loaded or enters standby. A voltage-controlled oscillator (VCO) monitors the feedback voltage and, when the latter crosses a certain level, the switching frequency linearly decreases down to 20 kHz as a very minimum. Below this value, acoustic noise could be an issue. When the load decreases further, the part reduces the duty ratio and enters skip-cycle. I have combined the VCO and a skip comparator in this controller (*Generic_CM_PWM_FF.asc*). The core of the part is similar to the CM version I already introduced:

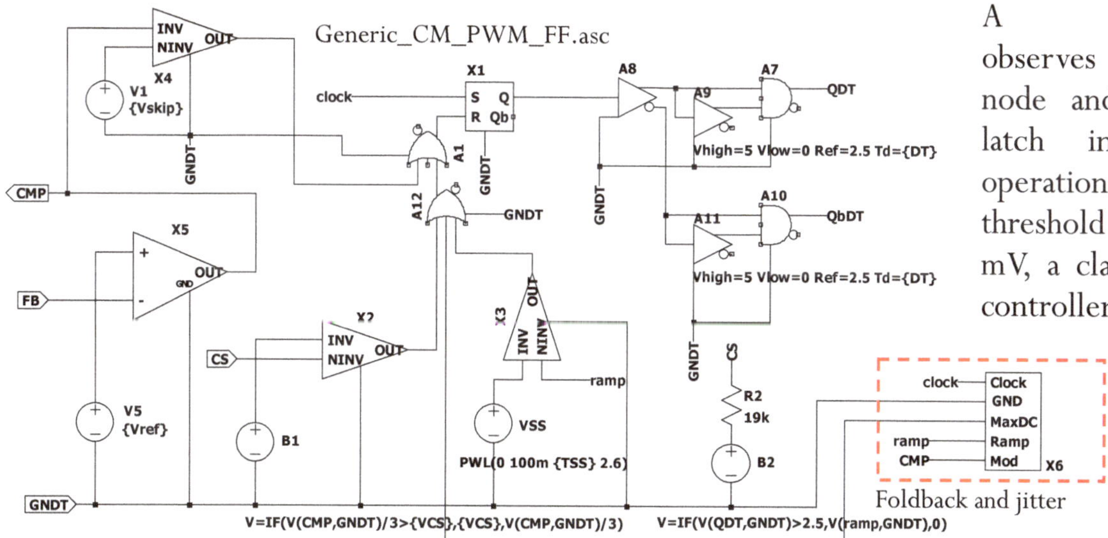

A comparator observes the CMP node and resets the latch in light-load operations. The skip threshold is set to 500 mV, a classic in these controllers.

The VCO block observes the feedback voltage V_{CMP}. When it is above 2 V, the frequency is constant. When V_{CMP} falls below this level, F_{sw} linearly reduces down to $V_{CMP} = 0.8$ V, at which point it is clamped to 20 kHz (all is programmable). The subcircuit includes jitter which modulates the frequency with a triangular shape.

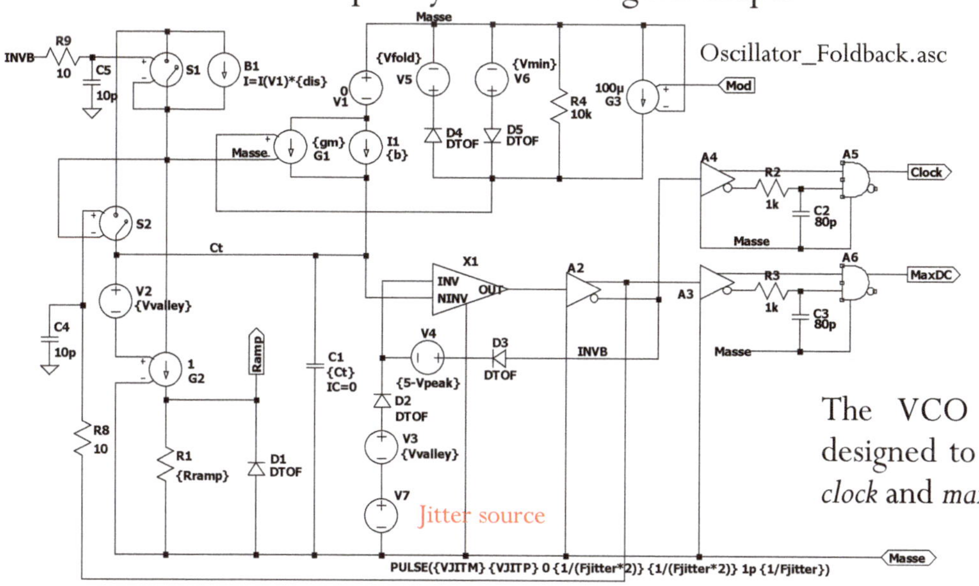


```
*
* Do not alter the below lines
*
.param Ct=100p
.param Vpeak=3
.param Vvalley=1
.param t1={maxDC/Fsw}
.param t2={(1-maxDC)/Fsw}
.param ichg={(Vpeak-Vvalley)*Ct/t1}
.param idis={(Vpeak-Vvalley)*Ct/t2}
.param t10={maxDC/Fswmin}
.param t20={(1-maxDC)/Fswmin}
.param ichg1={(Vpeak-Vvalley)*Ct/t10}
.param idis1={(Vpeak-Vvalley)*Ct/t20}
.param a={(ichg-ichg1)/(Vfold-Vmin)}
.param b={Ichg-a*Vfold}
.param gm={a}
.param dis={(1+(t1/t2))}
.param Rramp={ramp/(Vpeak-Vvalley)}
.param VJITM={-(Vvalley)*Depth/50}
.param VJITP={(Vvalley)*Depth/50}
.param AMP={VJITP-VJITM}
*
```

The VCO block is specifically designed to drive a latch, with its *clock* and *maxDC* signals.

Transient Simulation with Foldback

THE APPLICATION SCHEMATIC is a classical ac-dc adapter delivering 19 V/3.5 A from a universal input range. The file is *Flyback CM – FF – cbc.asc* and you change the input voltage or the load to check the action of the foldback. In the screenshot below, the load is nominal at a 330-V input voltage. The feedback is railing at 5 V at power up and the frequency is 65 kHz. When the controller breaks as V_{out} meets the target, the frequency reduces to 20 kHz and skip is entered. The feedback recovers and the frequency is now 45 kHz.

Flyback CM – FF - isolated - cbc.asc

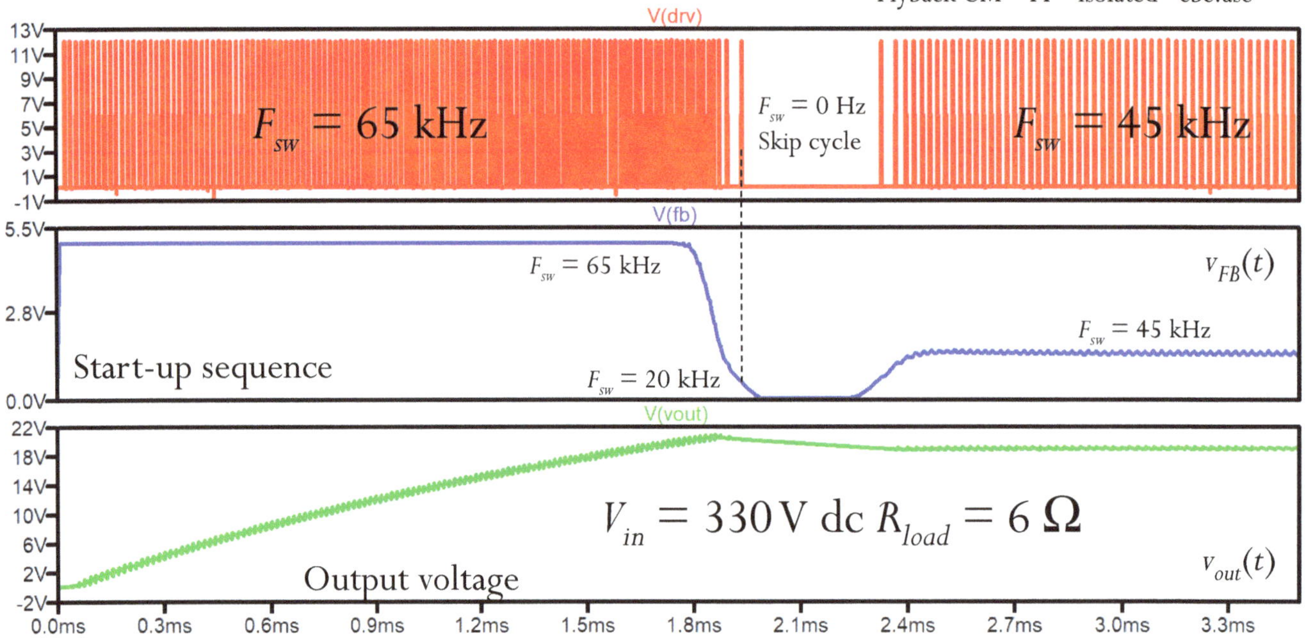

For a high input voltage – 370 V dc – or in no-load conditions, the frequency will hit the low limit of 20 kHz and will skip cycles, generating a bit of ripple on the output voltage.

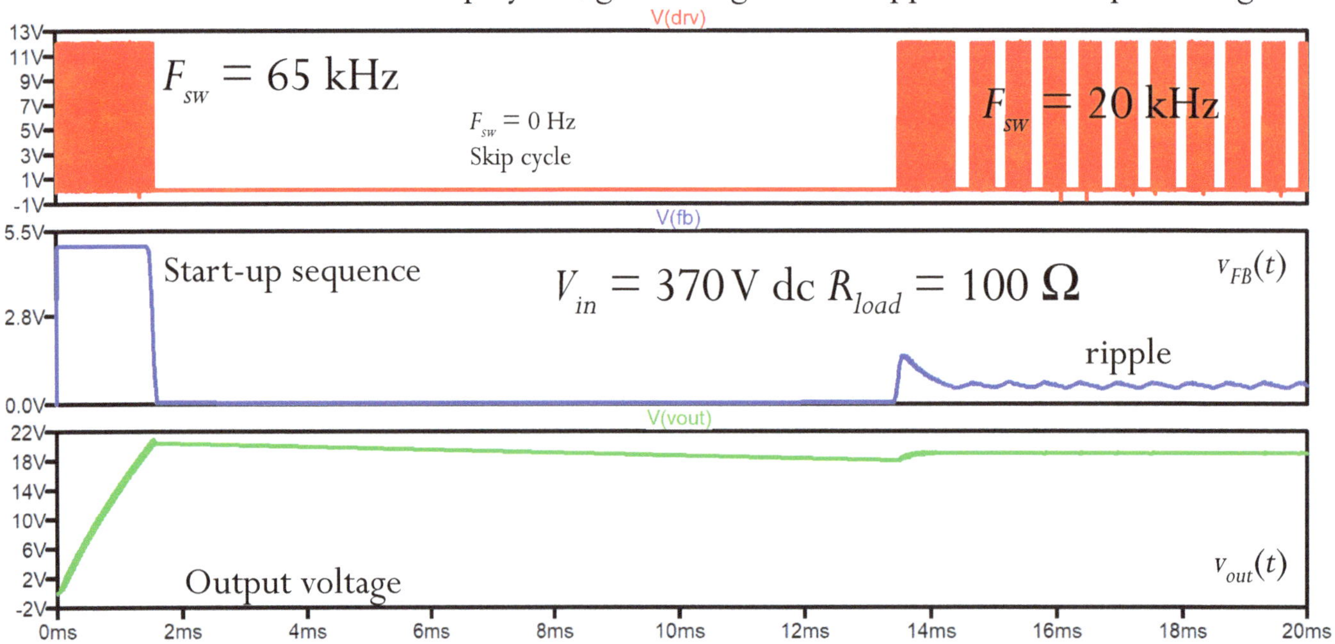

A Two-Switch Flyback Converter

THE LEAKAGE INDUCTANCE models the loose coupling between the primary and secondary sides of a transformer. In a classical flyback converter, this parasitic term stores energy during the on-time, which must be dissipated in heat, typically via an *RCD* clamp, after the switch has turned off. This extra loss affects efficiency and reliability of the converter. However, several options exist to recycle this energy – as with the active clamp flyback or the asymmetrical half-bridge – but the simplest of those, remains the two-switch flyback. By combining two power transistors and a pair of diodes, we can recycle the energy through the input source and improve efficiency. This is what the two-switch flyback offers. It has however been less successful than its forward converter cousin, probably because of the transformer core size that increases with power. Nevertheless, it is a robust converter, well suited for high-voltage applications.

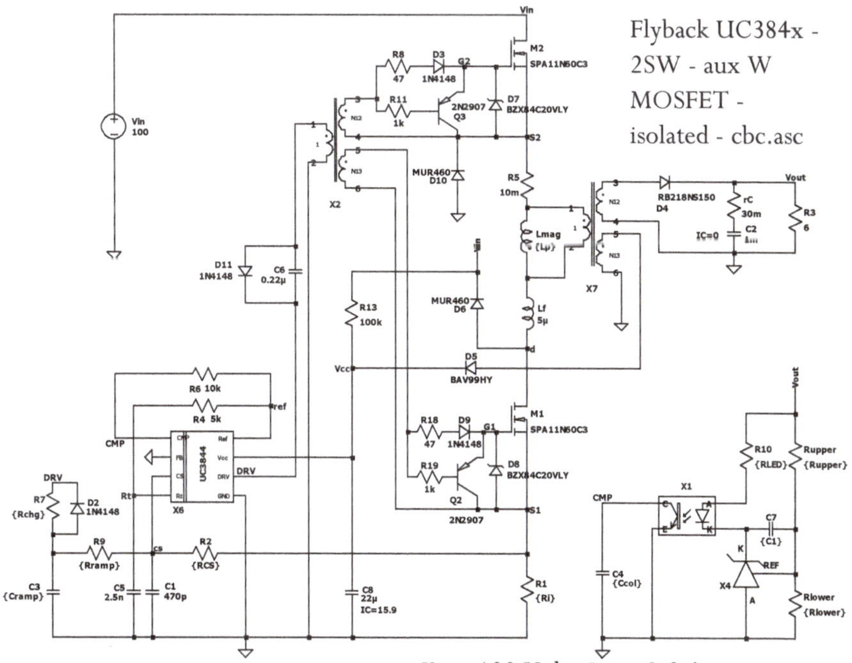

Flyback UC384x - 2SW - aux W MOSFET - isolated - cbc.asc

This configuration is very attractive when we think about recycling the leakage energy. Unfortunately, unlike with the forward converter where the magnetizing current is low, the current stress in this flyback converter is quite high at turn off. Considering the freewheel diodes D_5/D_{10} forward transit time, you can potentially still experience a significant voltage overshoot on the MOSFETs v_{DS}. This is true with low-C_{oss} types that favor high dV/dt: good layout and ultra-fast diodes are then required.

$V_{in} = 100$ V dc, $I_{out} = 3.2$ A

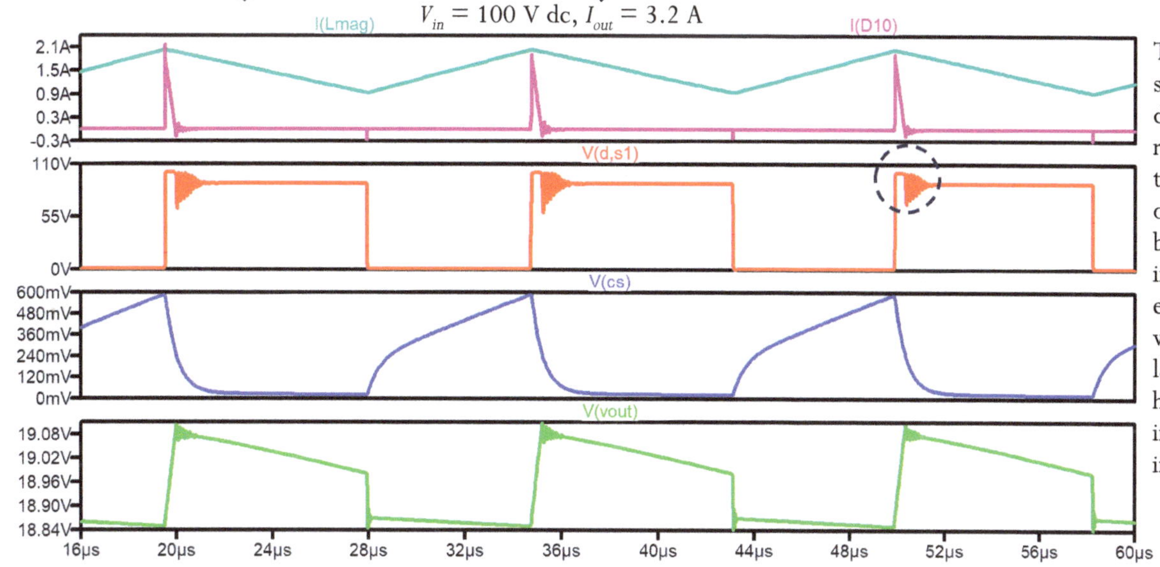

The simulation does not really show the extra overshoot, but it exists in reality, especially with a poor layout and high stray inductances in the path.

An Active-Clamp Flyback

THE ACTIVE-CLAMP FLYBACK (ACF) offers an attractive solution to entirely recycle the leakage energy and bring zero-voltage switching (ZVS). By eliminating the turn-on switching loss, the designer can increase the switching frequency, paving the path for high-density ac-dc or dc-dc converters. Several mechanisms exist to bring ZVS, but the basic principle is to impose a negative current in the magnetizing inductance, prior to releasing switch SW2 and before low-side switch SW1 is turned on again. This can be done in different ways, but here, we will adopt the typical option with a high-side transistor. The dead time, naturally present in our generic current-mode controller, will help in achieving our ZVS goal. The application schematic is shown below:

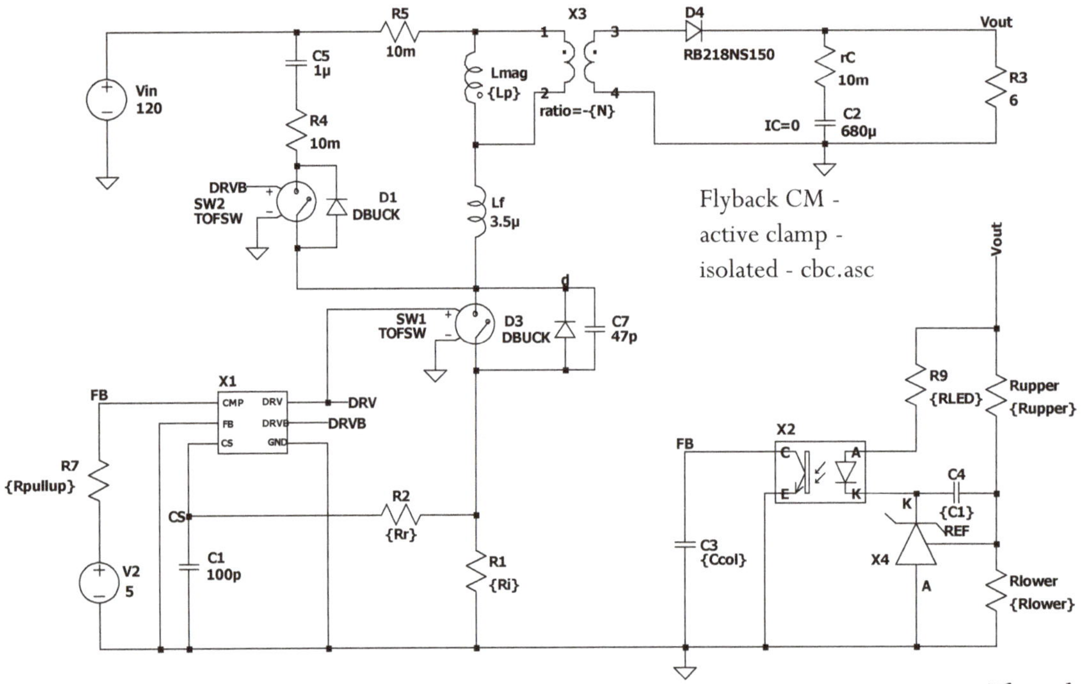

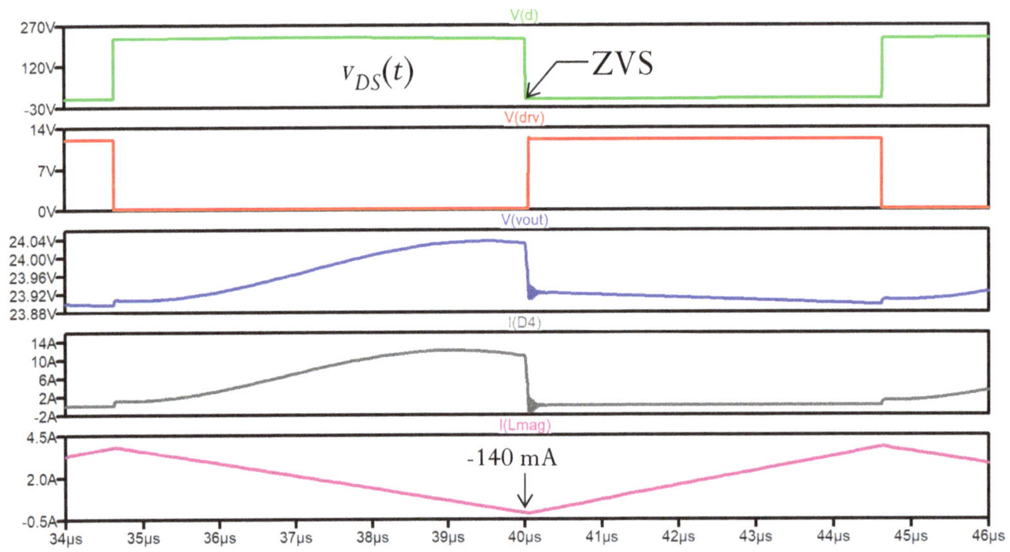

The plateau voltage during the off-time is almost flat for the v_{DS} waveform. The turn-on behavior of the secondary diode is very smooth. ZVS is confirmed on the drain:

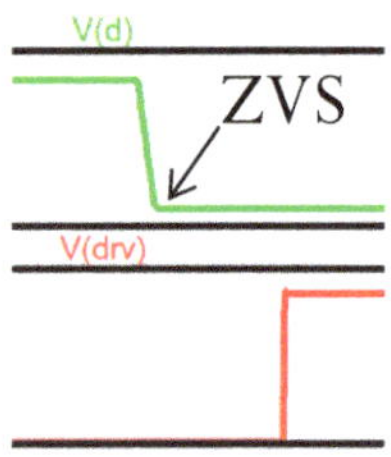

The Asymmetrical Half-Bridge

THE AHB CAN ALSO bring ZVS in the switching transistors, this time connected in a simple half-bridge configuration. In my opinion, this structure operates more safely, as the MOSFETs drain-source voltage cannot go higher than the dc bus level. The working principle differs from that of the ACF, but the operating waveforms are very similar. One option is to resort to a variable-frequency controller (QR) and regulate the negative magnetizing inductor current, adjusted to bring ZVS in any conditions.

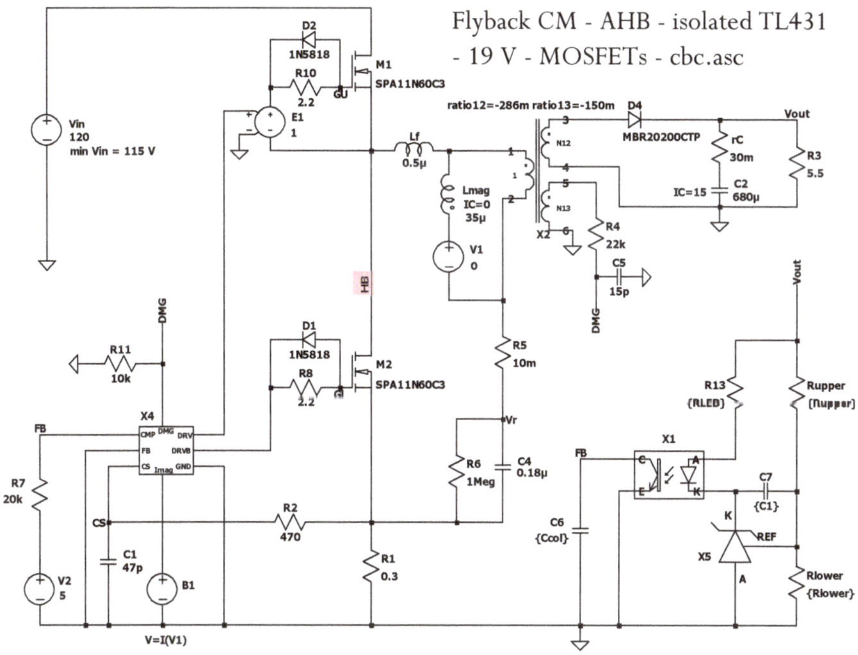

This modified QR controller hosts an extra comparator which programs the negative magnetizing current prior to actuating the high-side switch. The energy stored in the leakage inductance at that moment, should be sufficiently high to bring ZVS on the upper transistor. On the other hand, ZVS on the lower transistor is obtained more easily with a high peak current when the upper transistor turns off. Hence the different slopes on the half-bridge node.

The operating waveforms confirm the ZVS operation at both transitions. You can also observe the different transition times, linked to the different energy levels that are stored in the leakage inductance.

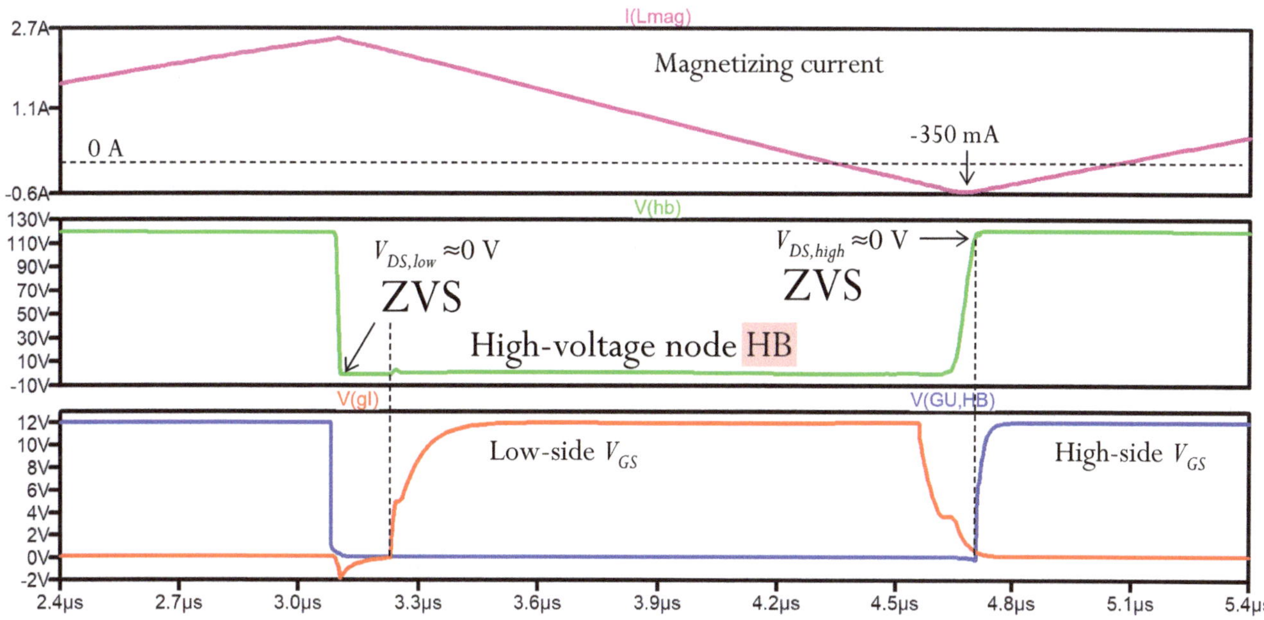

A Quasi-Resonant Flyback

THE QUASI-SQUARE-WAVE converter, also known as QR converter, is a popular structure mainly found in ac-dc power supplies. The controller implements a self-oscillating circuitry whose switching frequency depends on the line and load conditions. Operated in borderline conduction mode (BCM), the converter can meet ZVS or quasi-ZVS conditions when a small dead time is inserted prior to turning the primary MOSFET back on again. This way, when demagnetization is detected via a dedicated winding, the drain-source wave will freely swing to a minimum value – a valley. Turning the MOSFET on, right at this point, minimizes switching losses. So-called *valley-switching* operation still offers an efficient way to build high-density chargers, especially when coupled with a voltage-controlled oscillator (VCO) which folds the switching frequency back in light-load conditions, further improving efficiency. However, this simple controller does not host a VCO.

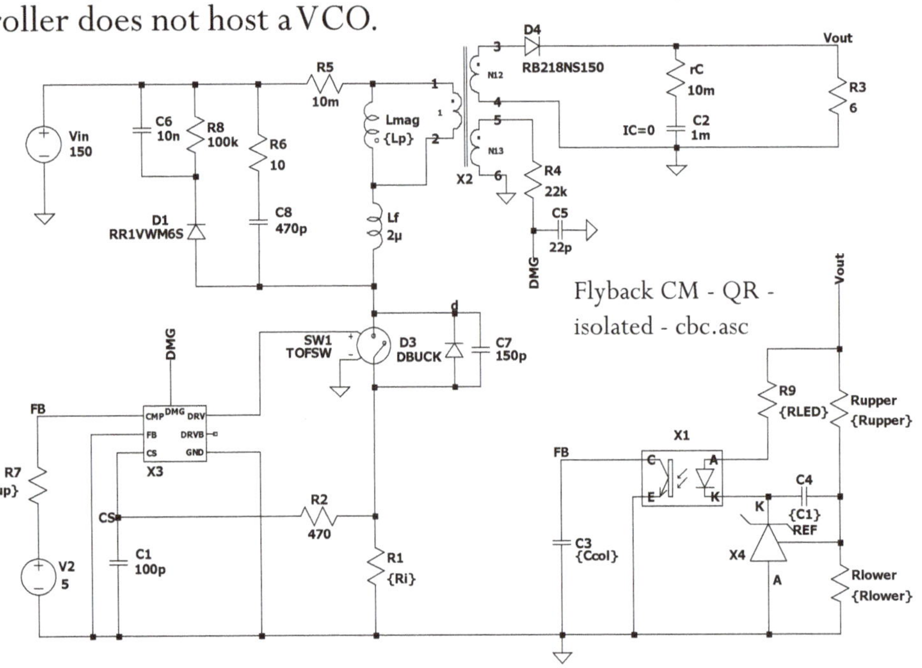

The auxiliary winding signal is delayed by the R_4C_5 time constant to exactly synchronize the turn-on event with the valley voltage. When quasi-ZVS is met, it is possible to increase C_7 across the drain-source and limit the turn-off dV/dt for less common-mode noise and radiated EMI.

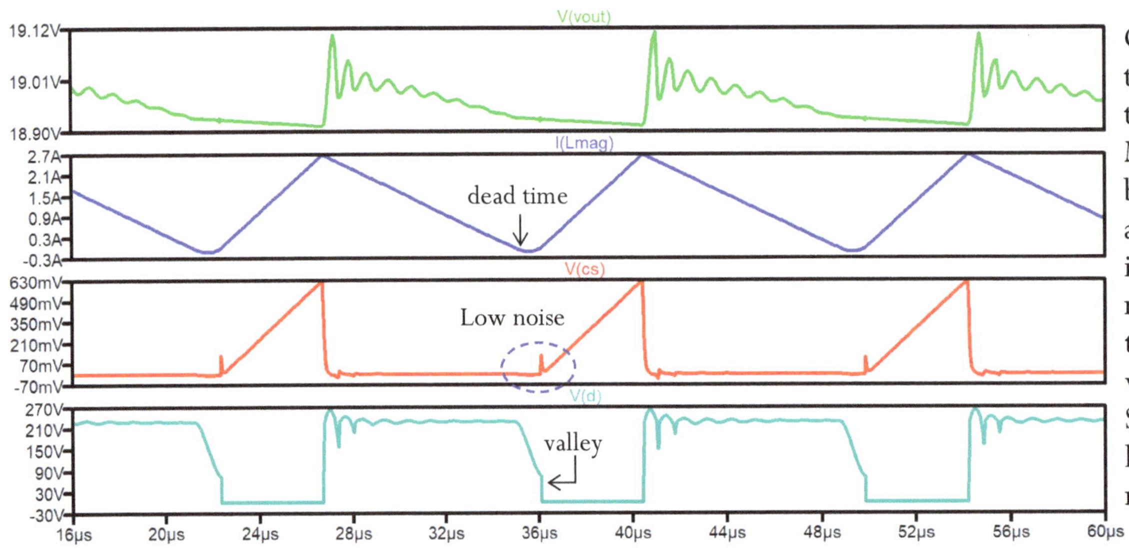

Good timing for turning the MOSFET back on again, right in the middle of the valley voltage. Switching losses are minimized.

A Multi-Output Flyback Converter

THE FLYBACK CONVERTER is often designed to deliver multiple output voltages, e.g. 5 V and 12 V in two separate rails. What matters in this case, is the transformer construction – how the windings are laid out – to ensure a good cross-regulation between the rails: if there is a load step on the 5-V output, I want my 12-V output to remain stable. Unfortunately, reflecting cross-regulation effects through transformer modeling, is extremely complicated as many inter-winding inductances need to be characterized. One good model to look at, is the extended cantilever model, described in *Modeling of Cross-Regulation in Multiple-Output Flyback Converters* by Bob Erickson and Dragan Maksimović. In this paper (there is also an APEC seminar with the same title), the authors describe a method to extract the parasitics of a multi-output transformer. Needless to say, the process is long and tedious but if you truly want to study regulation mechanisms in this type of configuration, this is a path to follow. To improve regulation, one known way is called *weighted* feedback: you assign a priority to a given rail that you want to see well regulated, while the second is less important but still monitored. For instance, 70% for the 5 V and 30% for the 12 V. The process consists of splitting the single upper sensing resistance into two resistances, respectively observing the 5- and 12-V outputs. The below average model shows how to configure the whole circuit.

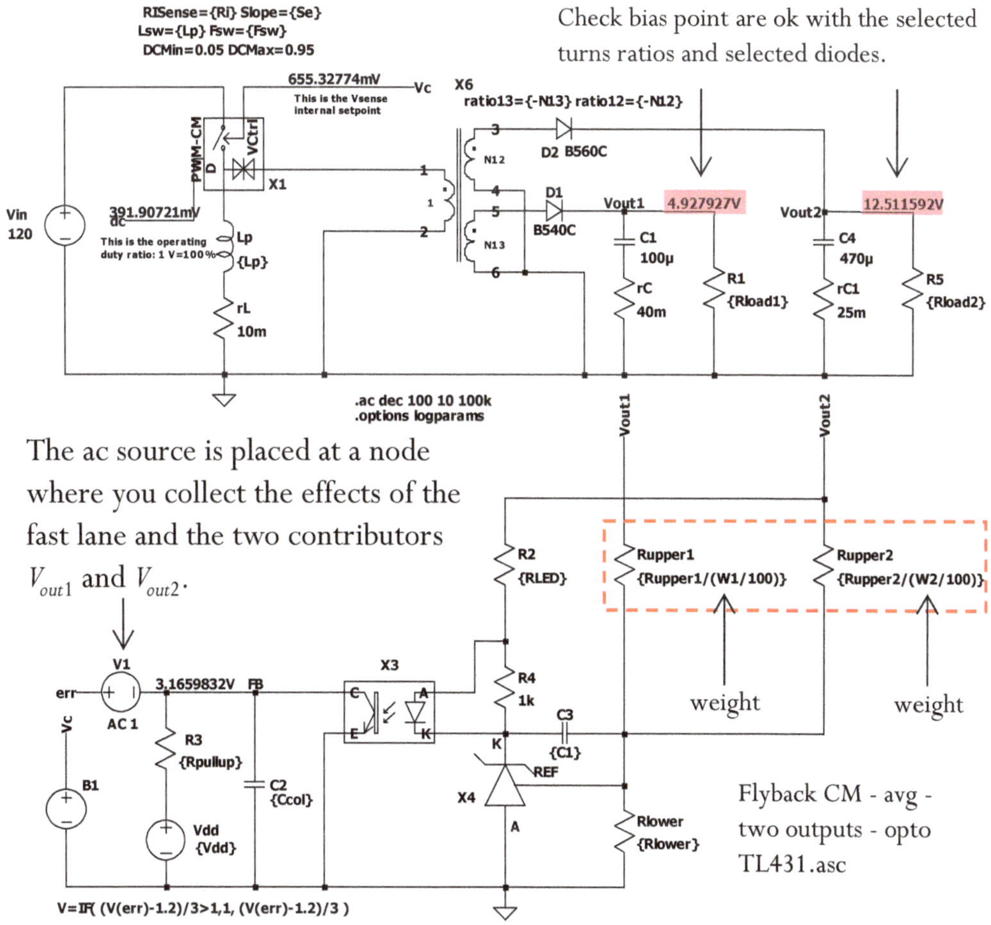

Each upper resistance is calculated based on the regulated voltage and its value is divided by the affected percentage. For instance, if R_{upper1} for the 5-V rail is calculated at 10 kΩ, once weighted by 70%, it will become 14.28 kΩ. Same thing for the second resistance, but weighted by 30% in this example.

A Multi-Output Flyback – AC Response

IN THIS PARTICULAR feedback configuration, extracting the control-to-output transfer function would be a difficult exercise because you have two outputs (or more with extra windings) merged at the TL431 REF node. Rather than applying superposition – this is what you can do when you have multiple feedback paths – I have placed the ac stimulus after the optocoupler, where the return lanes naturally merge. The diodes models are included and their forward voltage drops matter for determining the bias point at nominal loading. The turns ratios, in the end, are likely to be adjusted to obtain well-equilibrated outputs on the prototype. We will see in the cycle-by-cycle circuit, that I had to readjust the turns ratio of the 12-V rail. You could easily add another winding for modeling the auxiliary circuit if needed, it is easy with the generic transformer model.

When we run the ac analysis, the compensated response appears below:

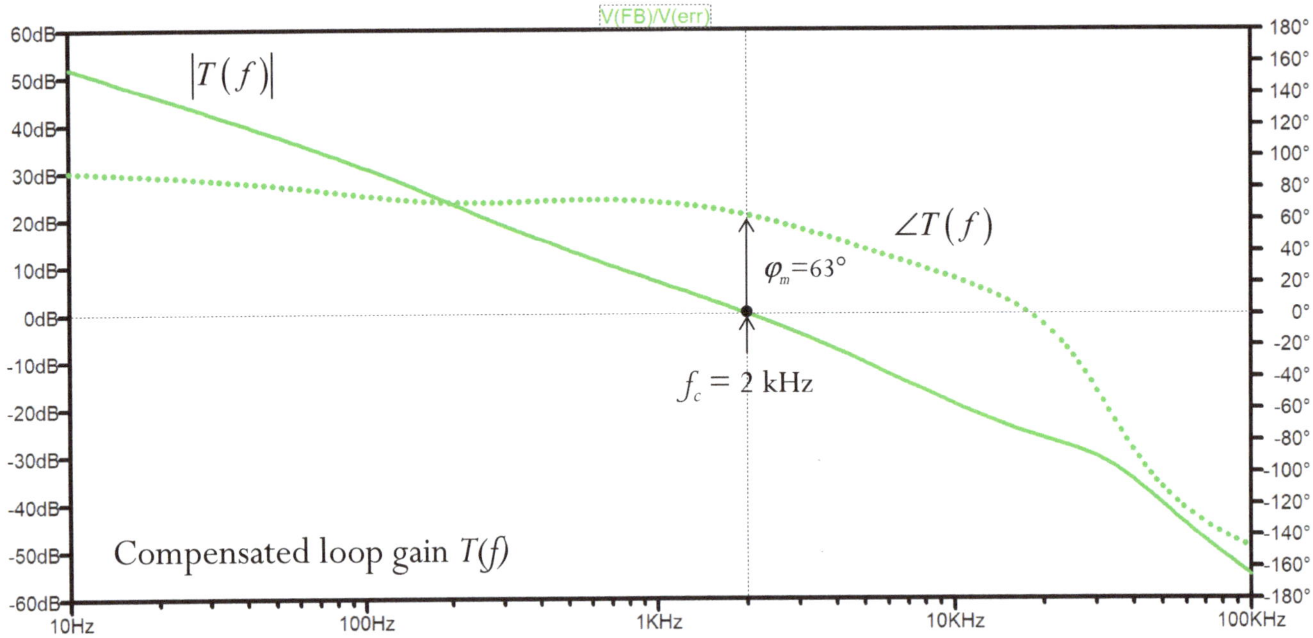

Without the control-to-output transfer function on hand, how do you enter *Gfc* and *PS* data in the macro? Provided you have a first ac plot of the loop gain, you observe the crossover frequency you have arbitrarily obtained. Assume you read 4.3 kHz versus the 2 kHz you set f_c at. Now read the magnitude at 2 kHz and you find it is 6 dB above the 0-dB axis. In this case, you will simply *increase* the gain passed in *Gfc* by 6 dB, forcing the compensator to reduce its mid-band gain by 6 dB. Assume *Gfc* was -14 dB, it becomes -14+6 = -8 dB and the curve goes down by 6 dB.

Now that crossover is correct, how do you increase or decrease phase margin? Either you change the *PM* parameter in the macro by the necessary amount, or you cheat the macro by *increasing* the *PS* parameter. Assume PS=-75° and you push it to PS=-85°, the compensator will compute the need for a 10° extra phase boost, naturally increasing the phase margin by these ten degrees.

Transient Simulations – I

RELIABLE TRANSIENT SIMULATIONS of a multi-output flyback converter are illusory, unless you have resorted to a good transformer model, such as the extended cantilever version. In this model, the multi-winding structure features all the inter-winding leakage inductances which delay the primary-side current at the switch opening. A possible version of a 4-winding model is shown below:

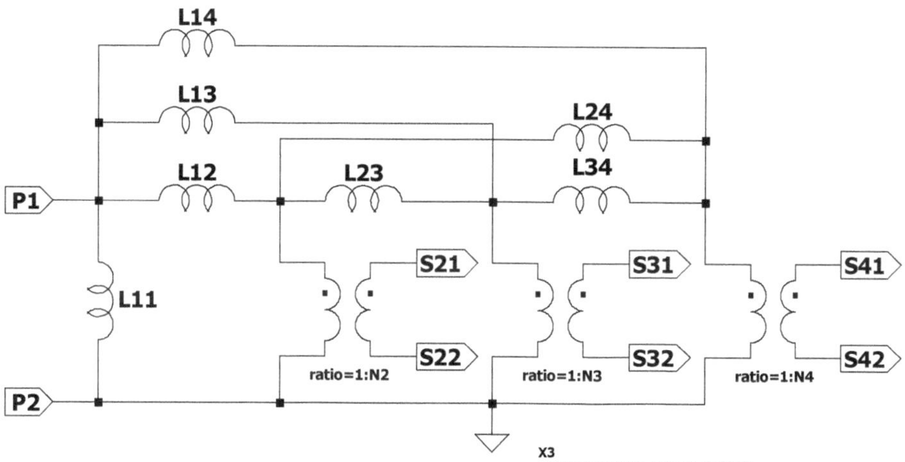

The many inductances in this model require a long extraction time, as you can imagine. However, this structure is interesting to realize how the primary current sharing between windings is hampered by the many inductances in its path.

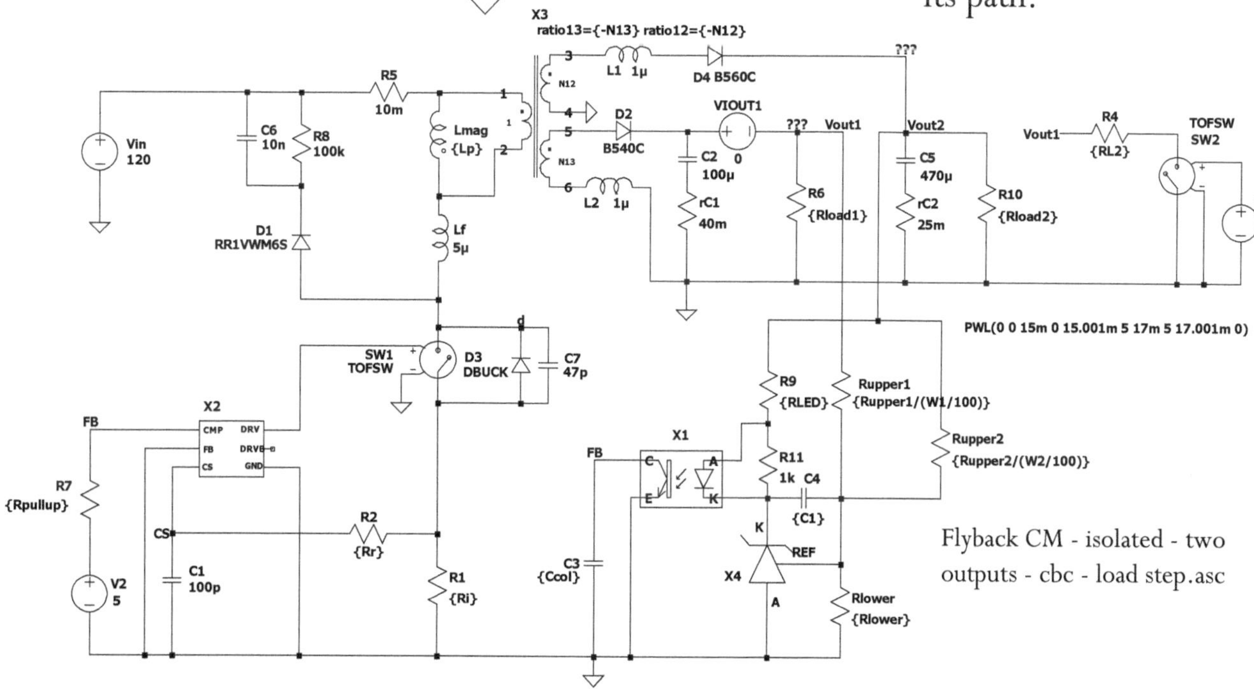

This is an overly simplified approach, in which I have added two leakage inductances in the secondary, like in a T-model. The extended cantilever subcircuit would require many more inductances, some of them being potentially negative (which may impact convergence). Also, keep in mind that the clamp voltage across R_8, plays a role in resetting the leakage inductance in the primary side. The diodes models are important here – the forward drop affects the operating point – especially if you add more unregulated windings.

Transient Simulations – II

THE REGULATION PERFORMANCE will heavily depend on the cross-winding inductances as we have seen. The first simulation result considers zero leakage inductances in the secondary side. For that purpose, I had to reduce the 5-V winding turns ratio to 55m. The regulation is excellent when I step the output current from 70% to 100% on the 5-V rail, the 12-V does not almost change:

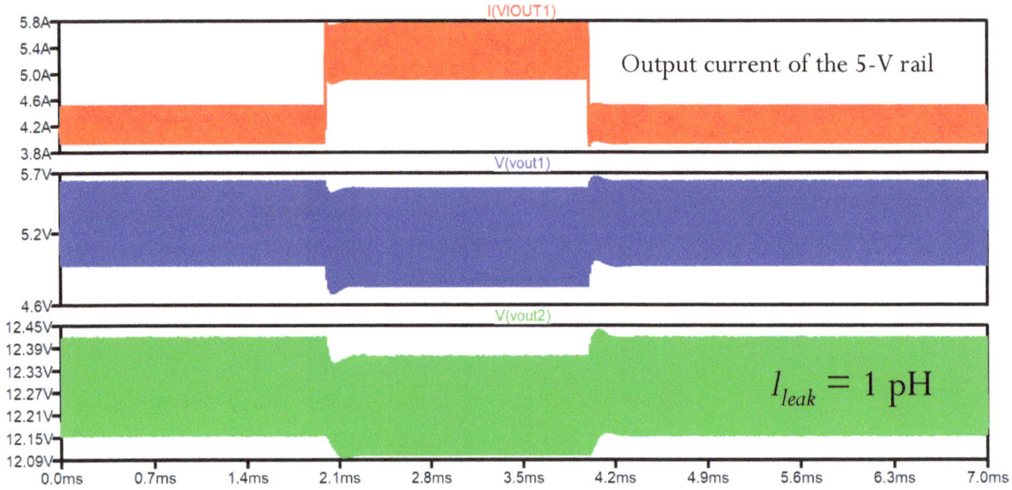

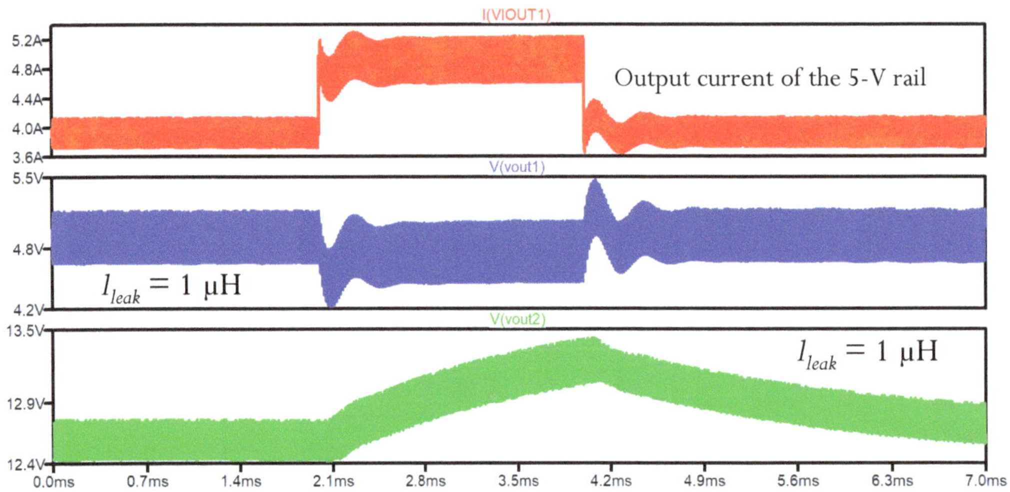

If I now increase the secondary-side leakage terms to 1 µH, I need to bring the 5-V turns ratio to 70m for an acceptable regulation point. However, the 12-V rail is now disturbed and increases significantly. You can imagine how difficult it is to bring the rails to the proper values on the bench, especially when multiple transformer design iterations are needed.

One possible option that I have seen in multi-output power supplies for printers, is the dc-stack of the windings:

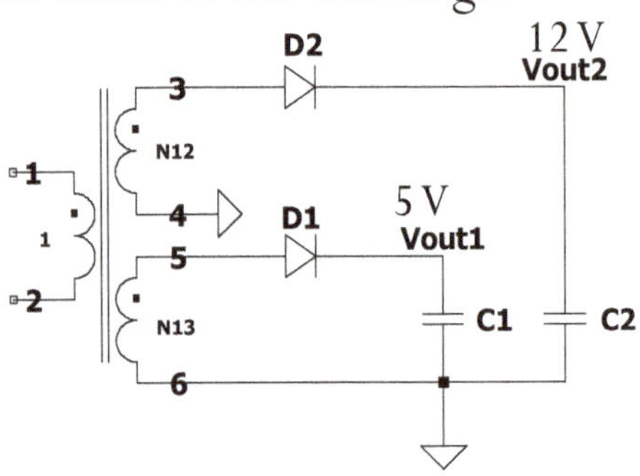

normal configuration

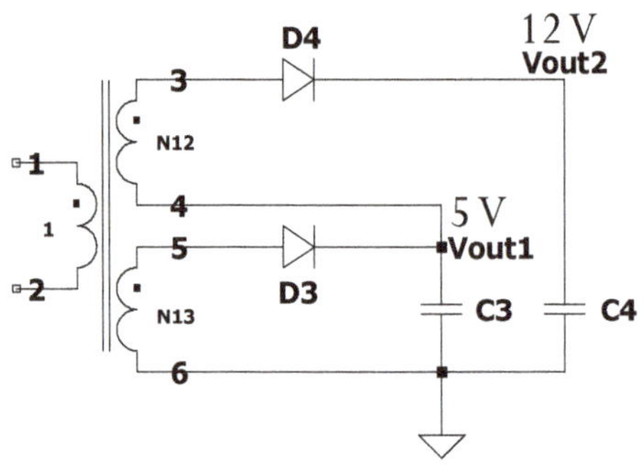

dc-stack

In this mode, the ground of the second winding, is connected to the 5-V rail. The turns ratio must be adjusted accordingly. More current now flows in the 5-V winding but cross-regulation is improved.

The SEPIC in CM

THE SINGLE-ENDED PRIMARY INDUCTOR CONVERTER, or SEPIC, offers the ability to boost or buck the input voltage, with a positive output polarity. In this example, the inductors are coupled and component values are taken from *SNVA168E* by Texas-Instruments. Current-mode control is a popular implementation, considering the distorted phase response obtained with the VM version. This *ANP135* document from Würth offers a comprehensive coverage of the SEPIC. Before attempting to run a cycle-by-cycle simulation, it is important to stabilize the loop with an averaged model. The CM PWM switch lends itself perfectly for this purpose. You will feed the model with the equivalent inductance value, which is the parallel combination of L_1 and L_2.

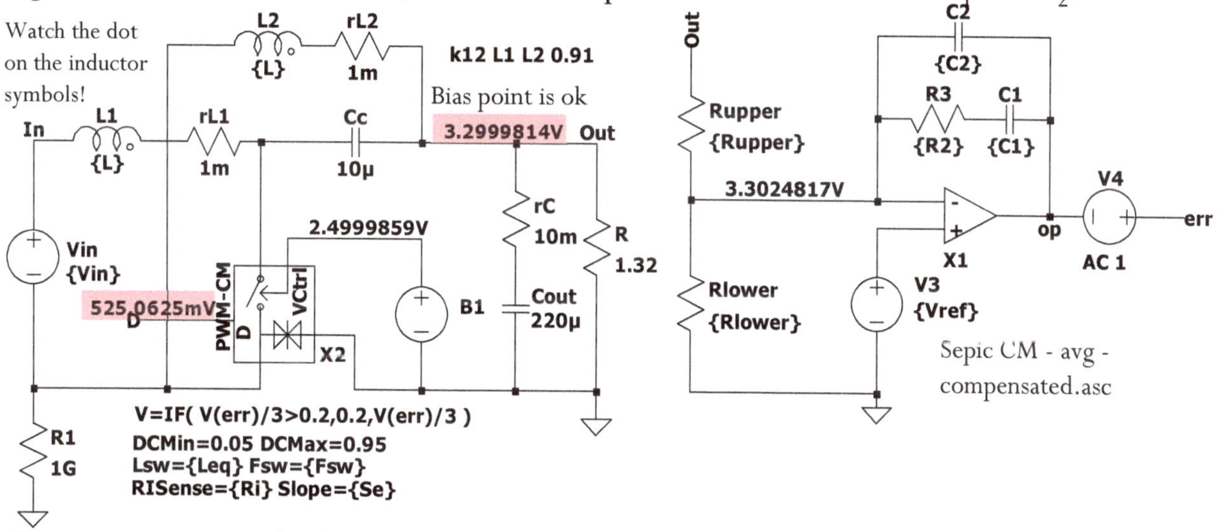

Once compensated, the converter exhibits a crossover frequency of 8 kHz, with a phase margin of 54°. The glitch can be attenuated by damping the series capacitor C_c, with an *RC* network placed in parallel.

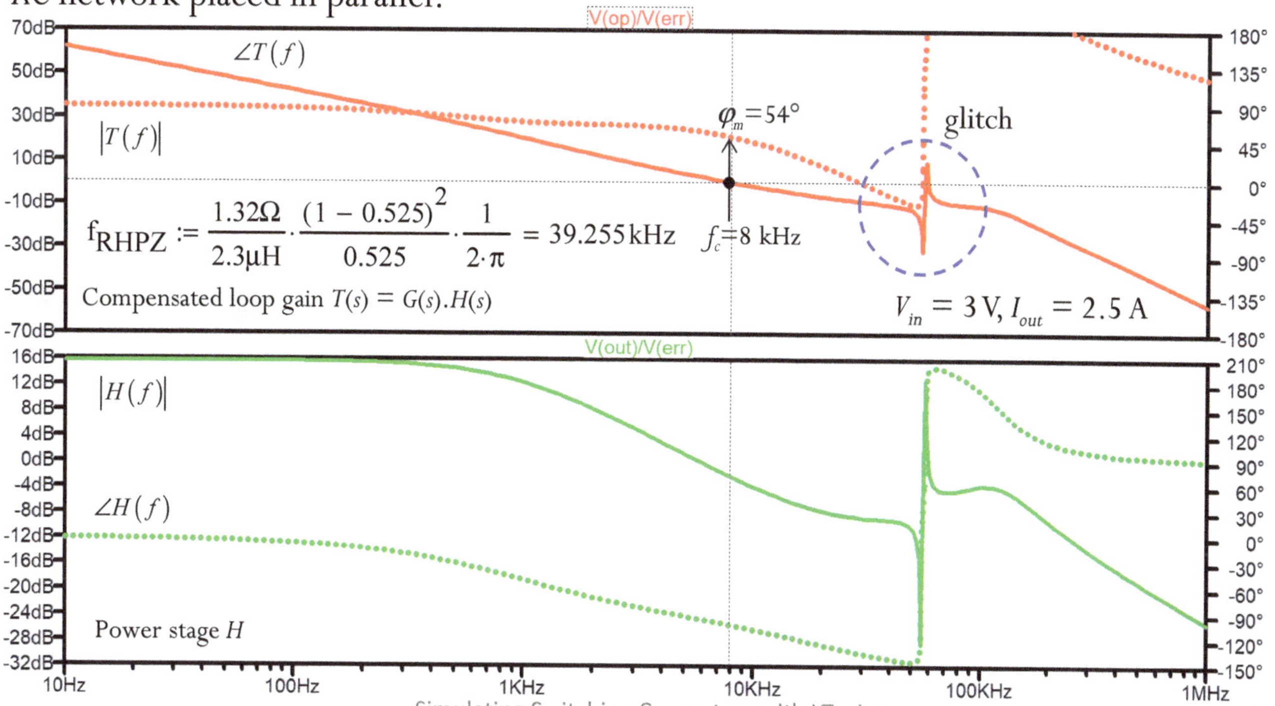

SEPIC CM Transient Simulation

NOW THAT THE SEPIC looks stable, we can put together a cycle-by-cycle circuit and check the transient response at both input levels, 3 and 5 V. I have used the generic CM model but did not wire the synchronous output for the sake of simplicity. The type 2 compensator is built around the internal op-amp. The 100-V MOSFET is obviously an overkill – *the only way to feel the noise* ♠ – and a lower BV_{DSS} type will also do the job. Check the V_{GS} at which the $r_{DS(on)}$ is specified though, considering a low input of 3 V.

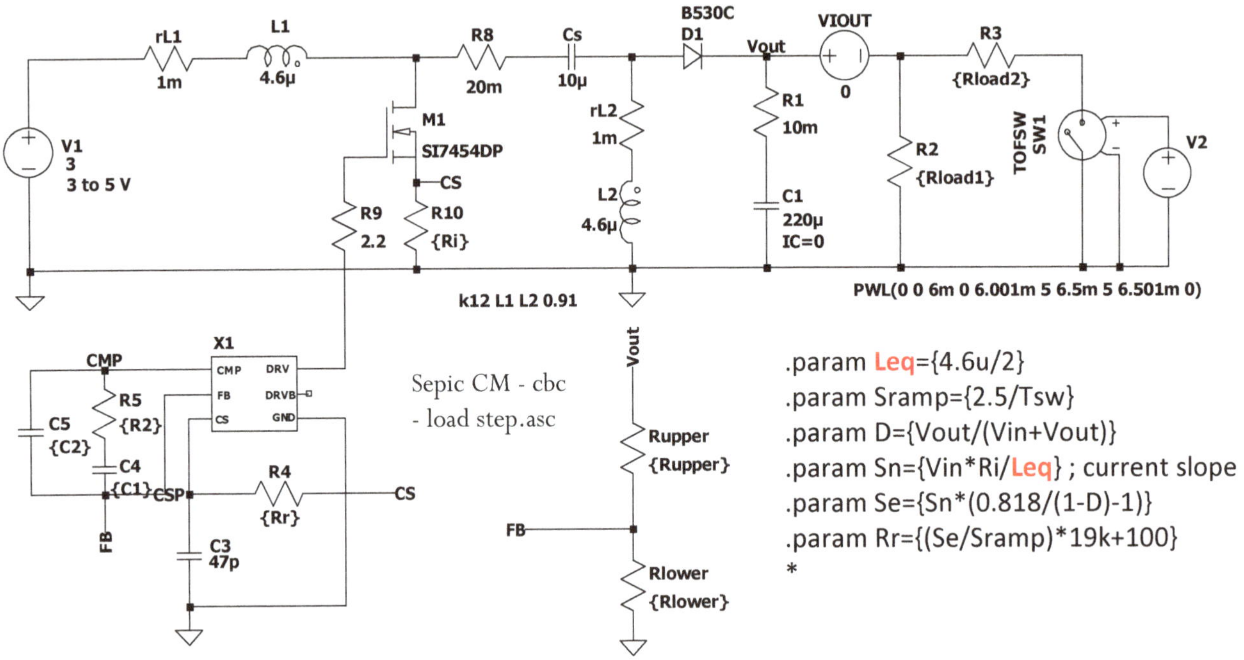

The transient response is excellent at 3 V and also at 5 V. The current is stepped from 1.25 A to 2.5 A with a high slope, considering the brutal closure of the switch. A PWL current source would let you impose the di/dt – e.g. 1 A/µs – and should be specified as a caption.

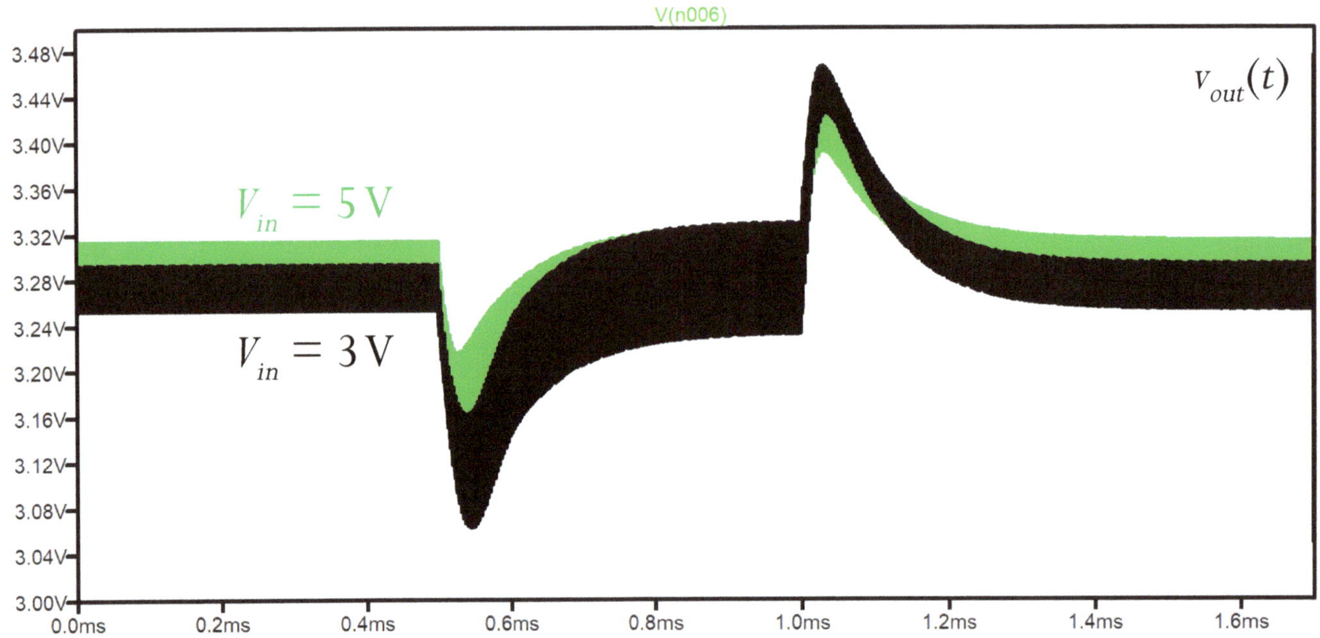

Averaged Ćuk VM Simulation

THE ĆUK CONVERTER was patented in 1977 by Dr. Slobodan M. Ćuk – pronounced *Chook* – and Dr. Robert D. Middlebrook (US patent 4,184,197). One important feature of this converter, is to cancel, in certain conditions, the input or output ripple current. Owing to its four energy-storing elements, it is a fourth-order converter. The PWM switch model is well suited for plotting the control-to-output transfer function of this non-isolated converter featuring coupled inductors. I have resorted to the transformer model in which leakage and magnetizing inductances are determined based on the adopted coupling coefficient. The circuit delivers 15 V / 3 A from a 10 to 20-V input.

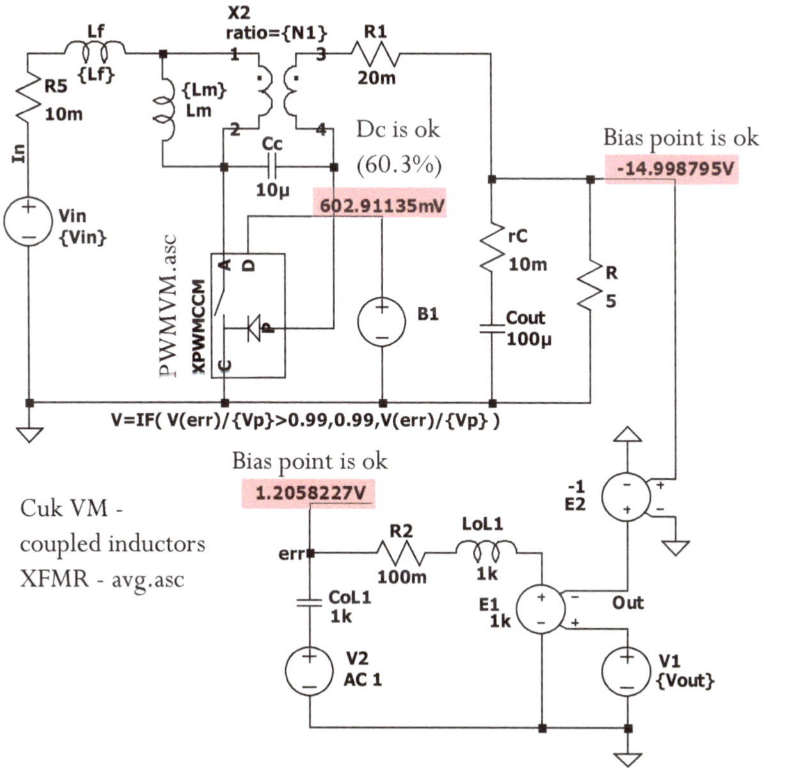


```
.param L1=20u ; first inductance
.param L2=50u ; second inductance
.param k=0.6325
.param Lf={L1*(1-k**2)} ; leakage
.param Lm={L1*k**2} ; magnetizing
.param N1={(1/k)*sqrt(L2/L1)}
*
.param Vin=10 ; input voltage
.param Vout=15 ; input voltage
.param Vp=2 ; this is the sawtooth
peak voltage
*
.options logparams
```

The ac response shows a very distorted phase plot. The crossover frequency must be carefully selected to stay away from the double pole occurring above 10 kHz. Its position depends on the coupling coefficient, the duty ratio and coupling capacitor C_c.

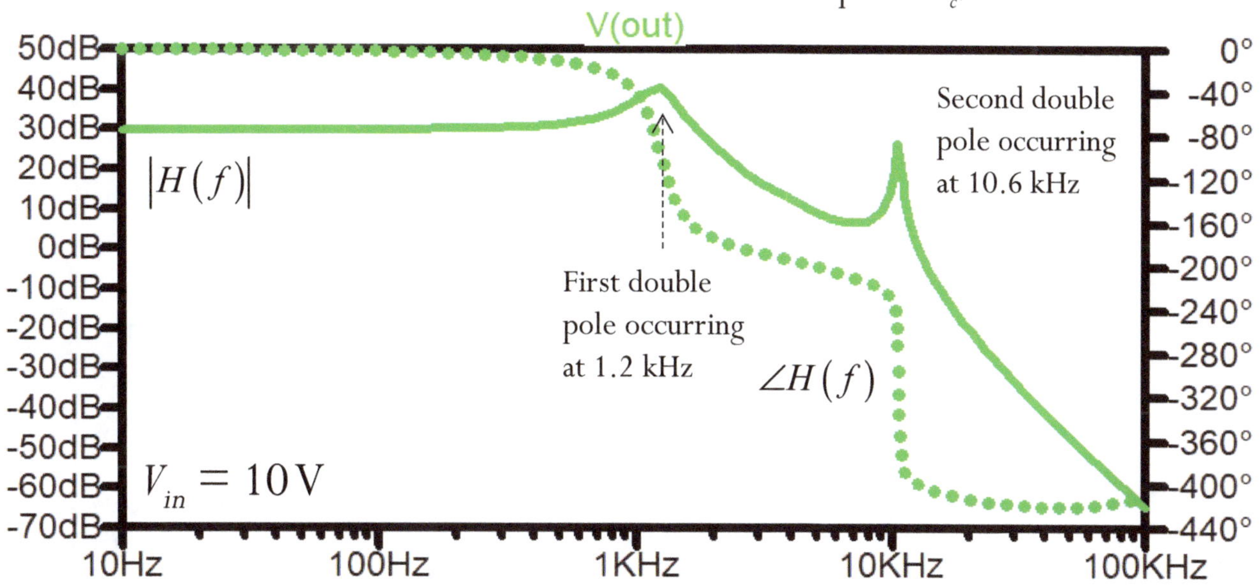

Averaged Zeta CM Simulation

THE ZETA CONVERTER is also a fourth-order converter, close to the SEPIC in architecture. It delivers a positive output whose value can be above or below the input voltage. In a SEPIC, the front-end stage is configured as a boost converter while it resembles that of a buck-boost in the Zeta, imposing a pulsating input current. However, the current flowing in the output capacitor is smooth, as with a buck power stage. The PWM switch can be easily applied to reveal the control-to-output transfer function of this converter.

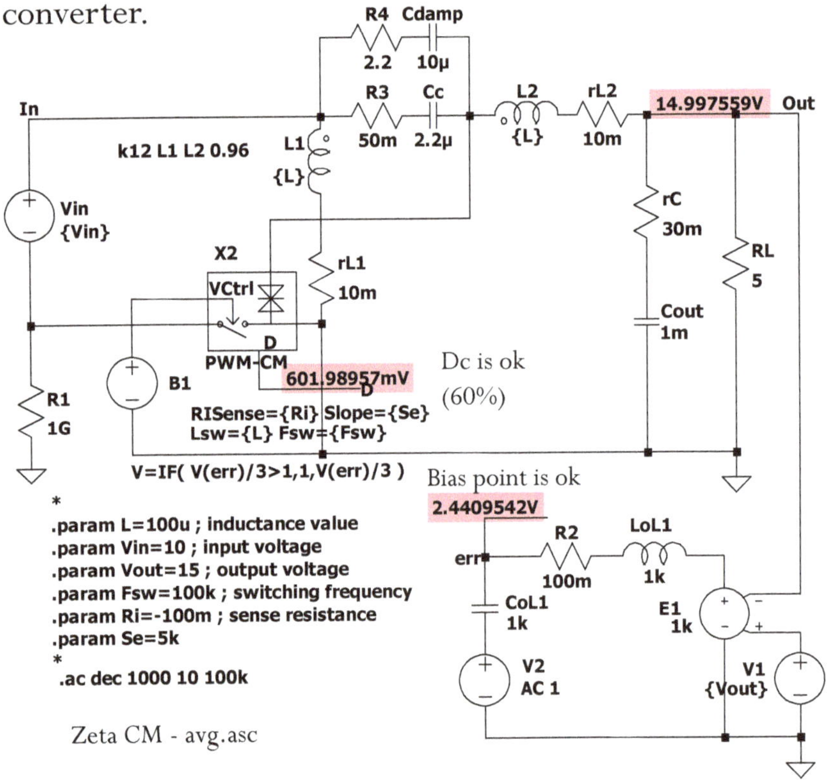

This circuit delivers 15 V/3 A from a 10-V input voltage. The duty ratio is set to 60% and a stabilizing ramp is added to damp the subharmonic poles. Please note the negative sense resistance value considering the way the model is arranged. In this mode, the input voltage is not ground-referenced but the control source and the output voltages are. The number of points in the ac analysis is set to 1000 considering the sharp resonance.

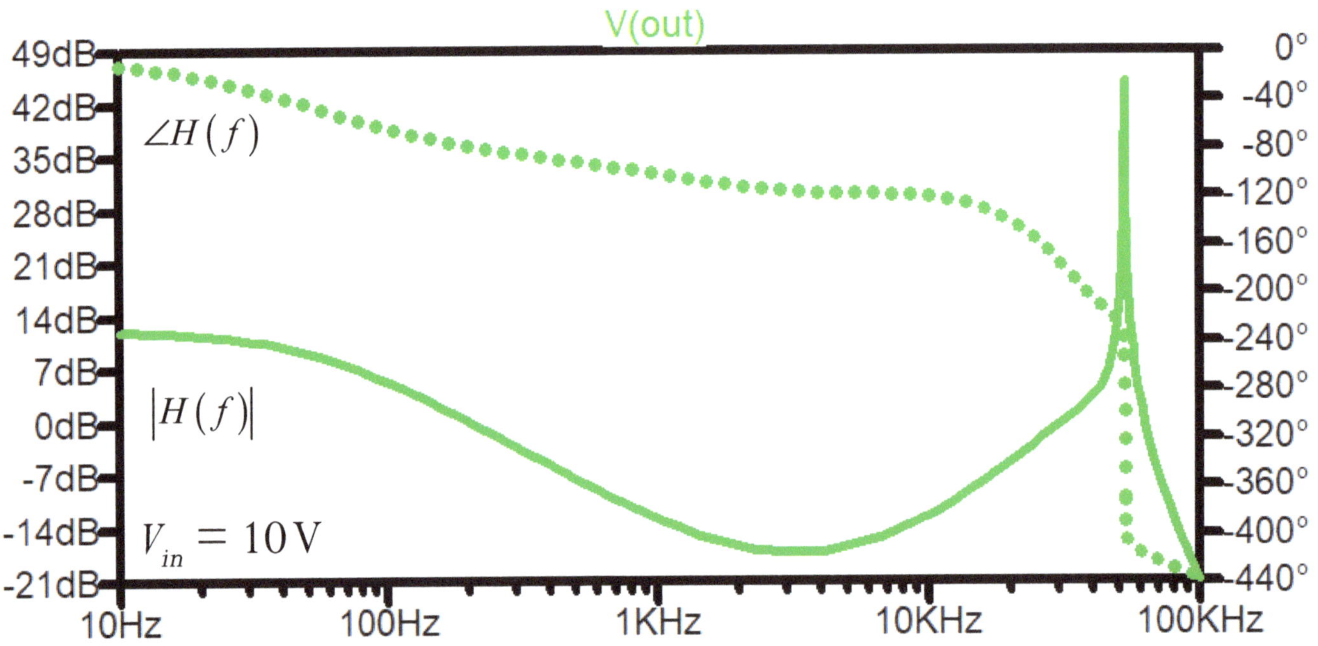

Averaged Zeta CM Transient Response

THE STUDY OF the transfer function reveals a distorted phase response which dips negatively as frequency approaches the resonant poles. Based on this observation, I have selected a crossover frequency below 1 kHz. Before going to the cycle-by-cycle analysis, it is simple and fast to confirm that the adopted compensation strategy leads to the expected response in presence of a load step. I am using a type 2 compensator here, and the voltage across the sense resistance is clamped to 1 V. The op-amp is the generic version I have already presented.

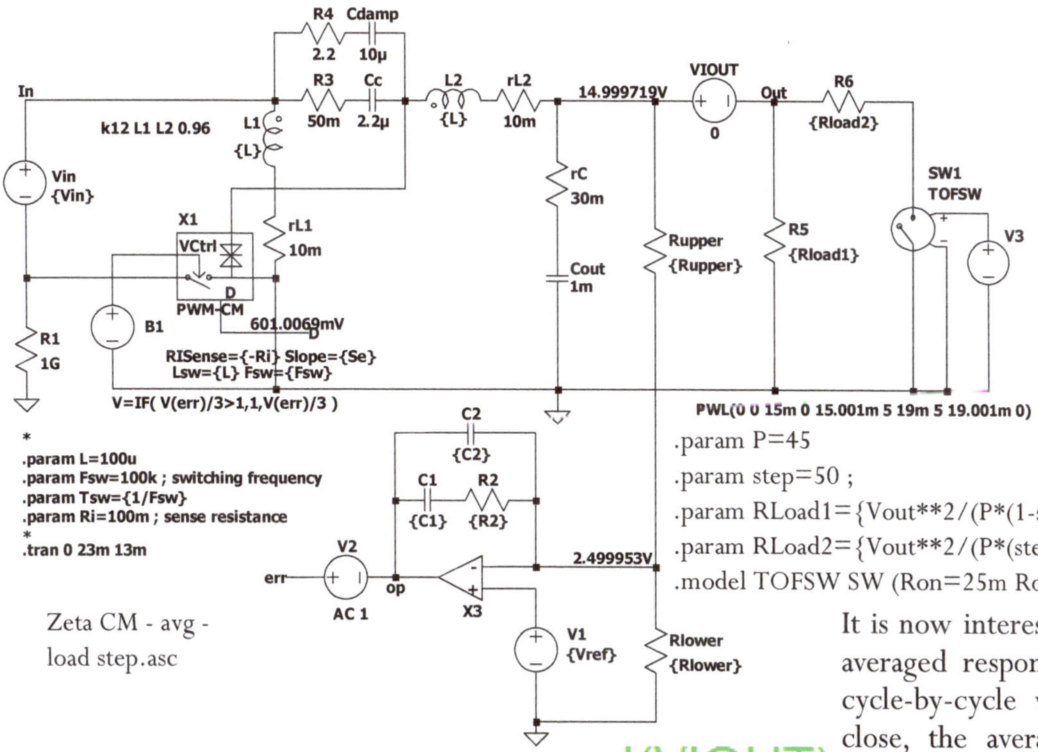

Zeta CM - avg - load step.asc

The current is stepped from 50 to 100% by paralleling two resistances via switch SW_1. The step duration is chosen for complete recovery before load release occurs. The result shows a stable output response, without significant undershoot (≈300 mV).

It is now interesting to compare the averaged response with that of the cycle-by-cycle version. If they are close, the averaged model can be trusted.

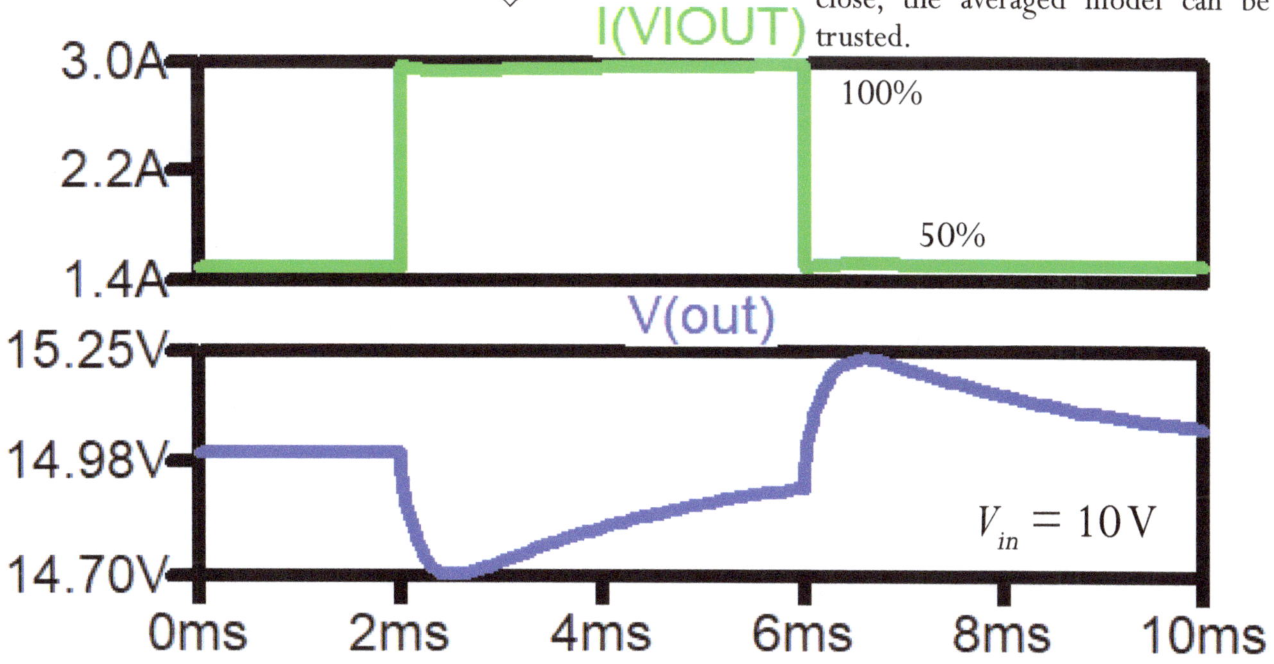

Zeta CM Cycle-by-Cycle

NOW THAT OUR AVERAGED model indicates a stable transient response, we can duplicate the compensation strategy and apply it to the cycle-by-cycle circuit. I have used my generic controller which lends itself well for simulating this type of converter. The DRVB output is used to drive a synchronous rectifier switch, helping for efficiency. The 2.2-µF coupling capacitor is supplemented by a damping network made of $R_6 C_{damp}$. Slope compensation is classically implemented via resistance R_4 whose value is automated by the macro, imposing a unity quality factor for the subharmonic poles.

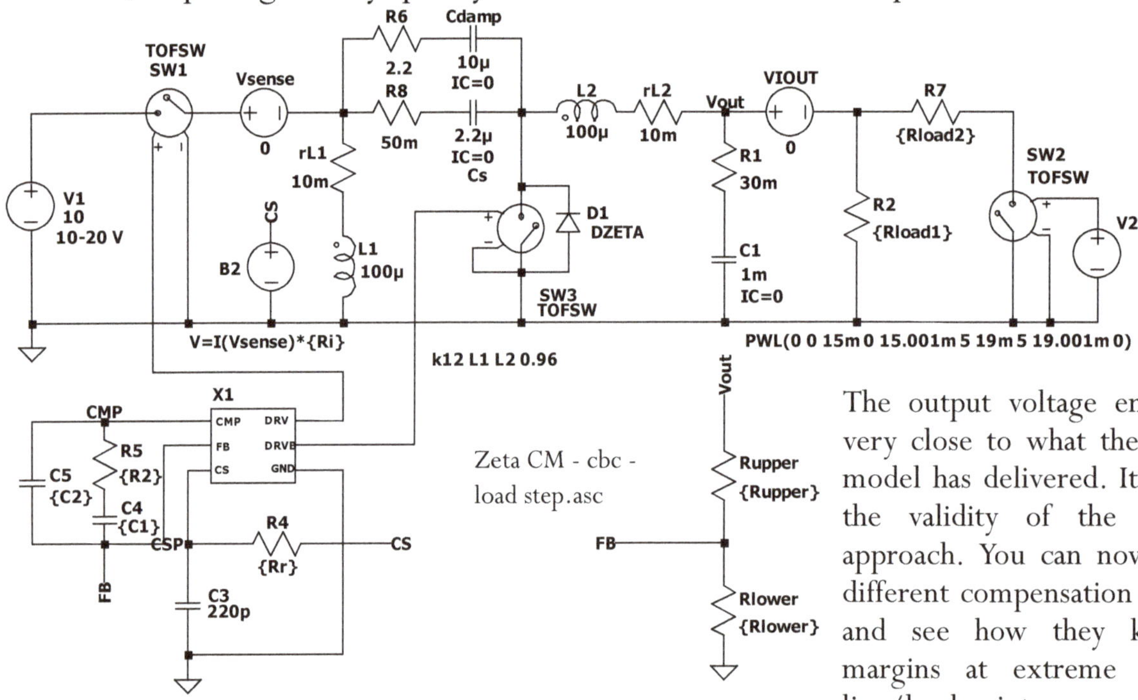

The output voltage envelope is very close to what the averaged model has delivered. It confirms the validity of the modeling approach. You can now explore different compensation strategies and see how they keep safe margins at extreme operating line/load points.

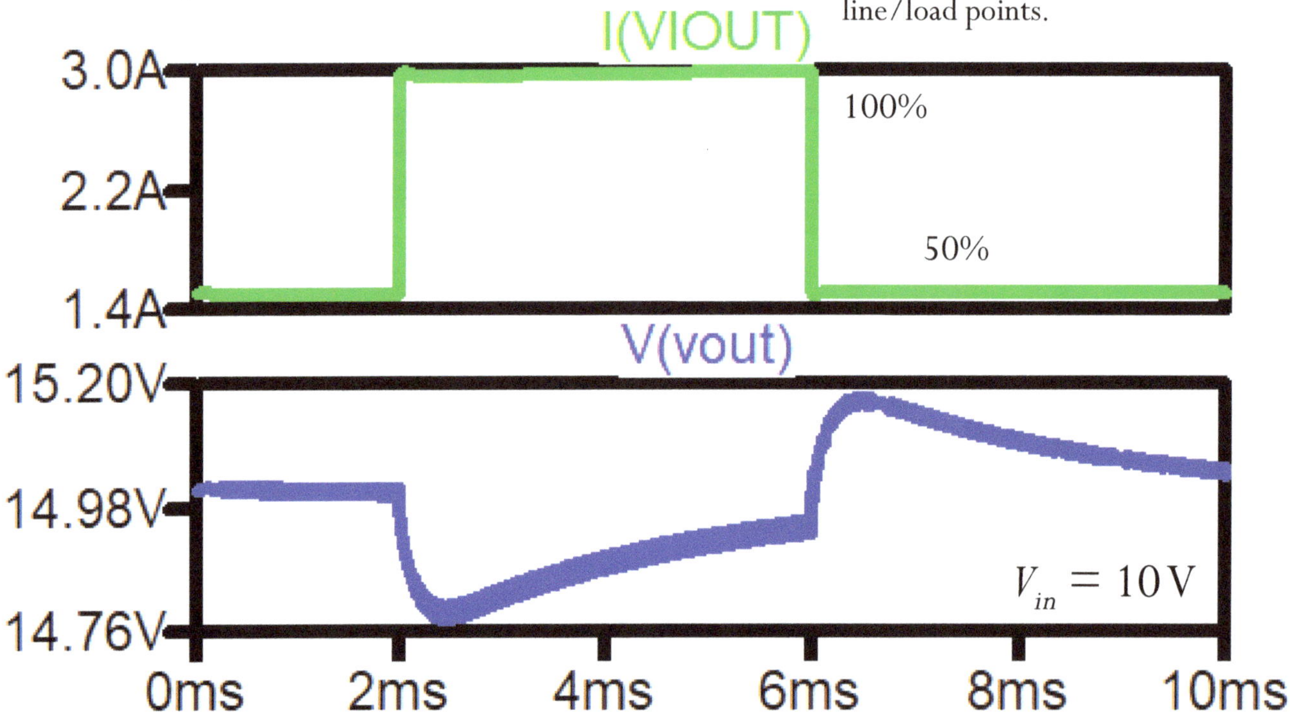

DC Transfer Characteristic of the LLC - 1

WHEN YOU DESIGN an LLC converter, it is interesting to graph its dc transfer characteristic. For the selection of the three energy-storing components, the variation of V_{out} versus switching frequency must be studied. You can certainly plot these curves using an Excel sheet or a mathematical solver, but I remember building a SPICE dc model when developing the NCP1395, many years ago. I have ported it to LTspice so that you can also run these simulations.

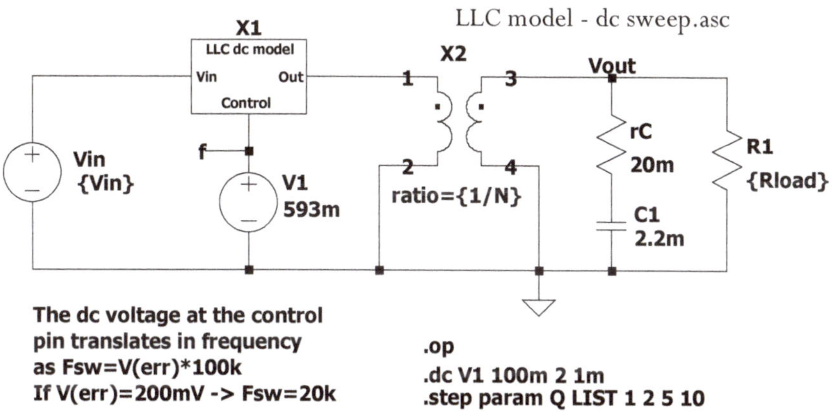


```
*
.param Vin=380
.param Vout=24
.param Pout=250
.param Fmin=40k
.param Fmax=250k
.param Ratio=2 ; choose ratio Lm over Ls
.param Lm=600u
.param Ls={Lm/Ratio}
.param Cs=33n
.param N=8
.param Z0={sqrt(Ls/Cs)}
*
.param Q=2
.param Rload={Q*Z0/N**2}
.param Eta=0.9
.param Rac={8*Rload*N**2/(pi**2*Eta)}
.param Fs={1/(2*pi*sqrt(Ls*Cs))}
.param Fm={1/(2*pi*sqrt((Ls+Lm)*Cs))}
*
.nodeset V(Vout)={Vout}
*
```

The right-side macro lets you select the LLC parameters such as the load and the minimum/maximum switching frequencies. The VCO outputs a frequency of 100 kHz for a 1-V input. The 593-mV bias thus implies a 59.3-kHz switching frequency. The dc sweep varies this voltage between 100 mV (10 kHz) up to 2 V (200 kHz) while different quality factors (hence loads) are explored.

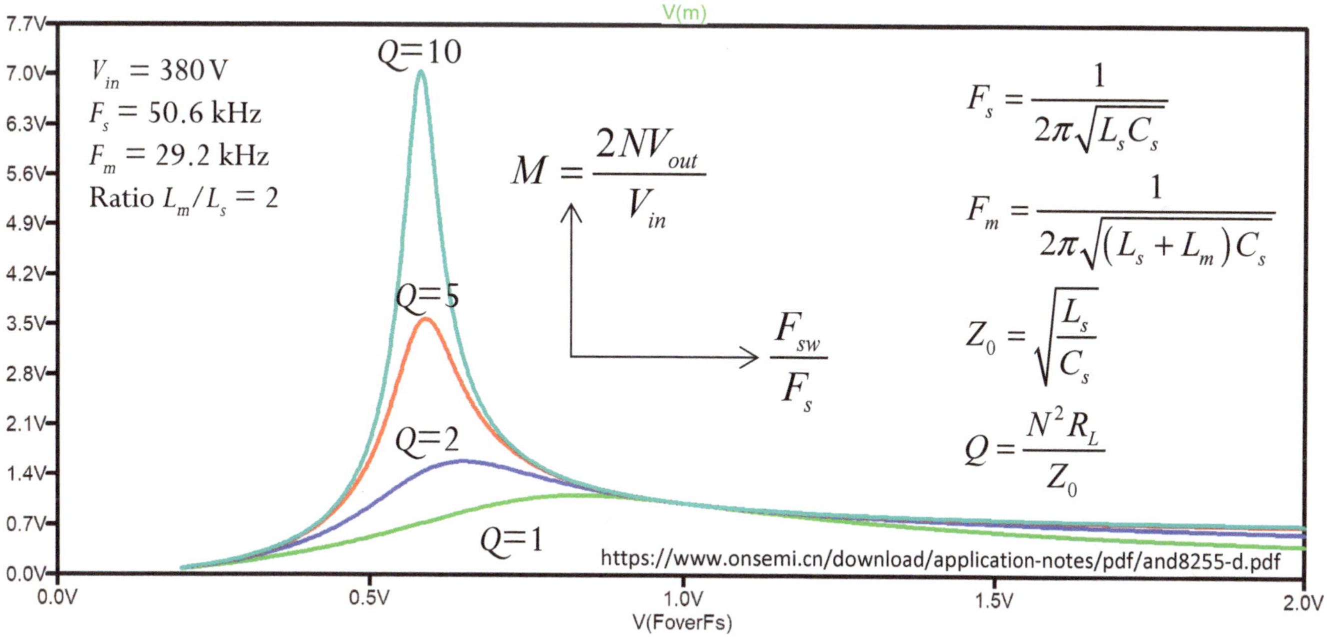

DC Transfer Characteristic of the LLC - 2

THE DC MODEL shown in the previous page, can sometimes have difficulty to find its operating point. A simpler and more robust approach, consists of capturing an expression in which the parameters you sweep are independent from the resonant component values. For instance, L_m/L_s is a ratio that can be chosen to be 2, 4, or 6 for instance. Same for Q that we can sweep from 0.5 to 10 without referring to a specific part. This is what the below formula shows:

$$M = \frac{2NV_{out}}{V_{in}} \approx \frac{1}{\left| 1 + \frac{1}{L_N}\left[1 - \left(\frac{1}{f_N}\right)^2\right] + j\frac{\pi^2}{8}\frac{1}{Q}\left(f_N - \frac{1}{f_N}\right) \right|} \qquad L_N = \frac{L_m}{L_s} \qquad f_N = \frac{F_{sw}}{F_s}$$

Ratio between magnetizing and resonant inductances.

Ratio between switching and resonant frequencies.

M. Jovanović, *Principle of Resonant Power Conversion*, Delta in-house course, 2004

Extract magnitude

$$V_{out} \approx \frac{V_{in}}{2N}\frac{1}{\sqrt{\left(1 + \frac{1}{L_N}\left[1 - \left(\frac{1}{f_N}\right)^2\right]\right)^2 + \left(\frac{\pi^2}{8Q}\left(f_N - \frac{1}{f_N}\right)\right)^2}}$$

The LTspice circuit consists of source B_1 which models the above expression, plus source V_1 which corresponds to f_N and mimics the frequency sweeps. The ratio M is plotted on the vertical axis and shows the classical set of curves obtained for an LLC. Pay attention, N, in this expression, is N_p/N_s.

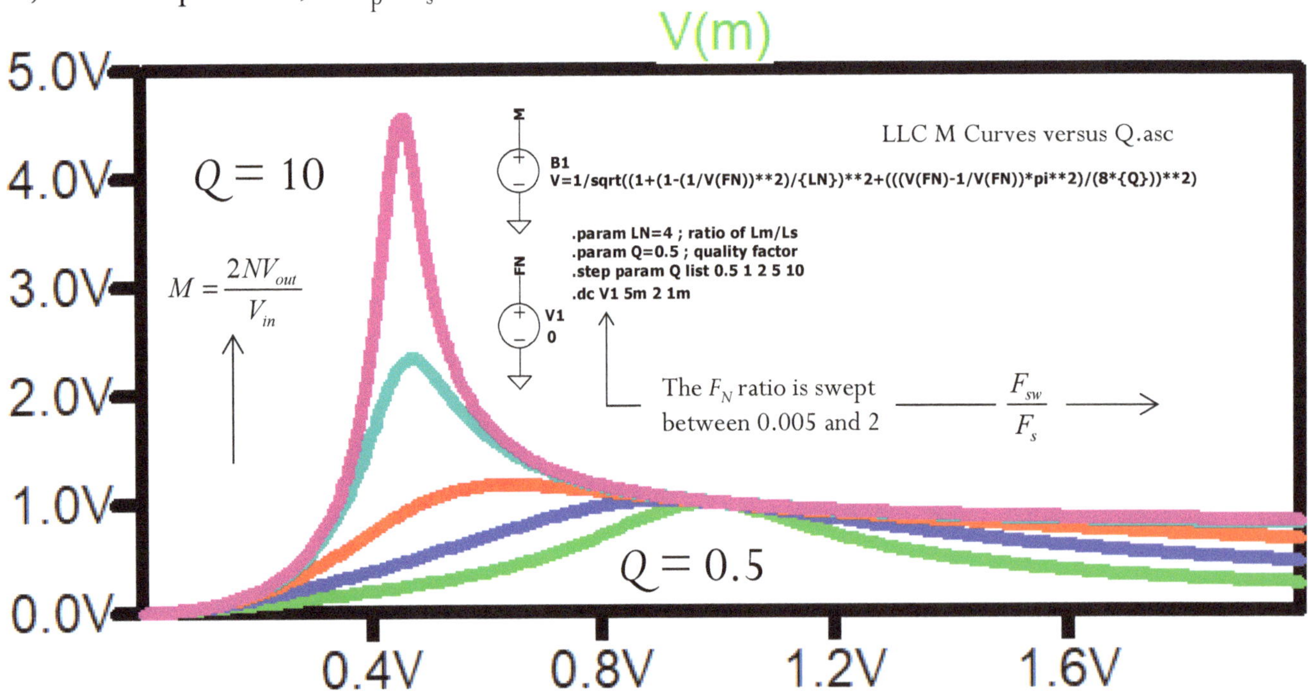

Open-Loop LLC Converter

THE LLC CONVERTER is built around three energy-storing elements: two inductors and one capacitor. The transformer magnetizing inductance L_m and the leakage element L_s (or an additional series inductor), form a resonant network with the series capacitor C_s. By adjusting the switching frequency via a voltage-controlled oscillator (VCO), it becomes possible to control the power flow. Different switching characteristics are obtained whether you drive the LLC below, at or above the resonant frequency. Before going into a complicated simulation at once, I recommend to test the calculated components values with a simple open-loop configuration, from which you can verify if the waveforms meet your criteria: ZVS in the primary, zero-current switching (ZCS) in the secondary, rms currents in the switches, the transformer, magnetizing current amplitude, and so on. When all your requirements are obtained, you can copy the resonant circuit and run a more complex simulation with a controller.

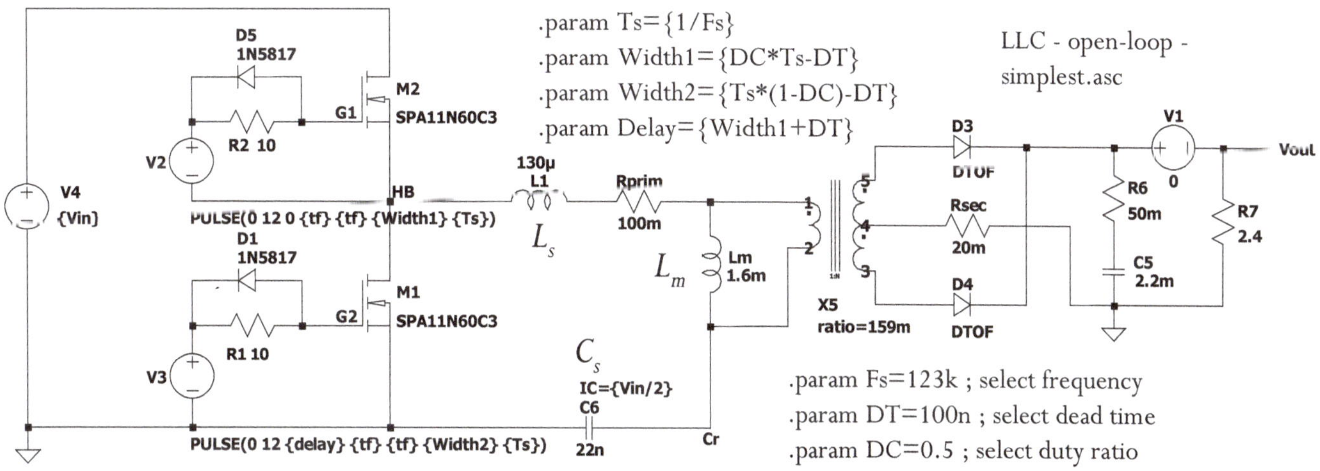

The circuit operates above resonance in this particular 24-V/10-A example. The input voltage is 390 V dc with a switching frequency of 123 kHz.

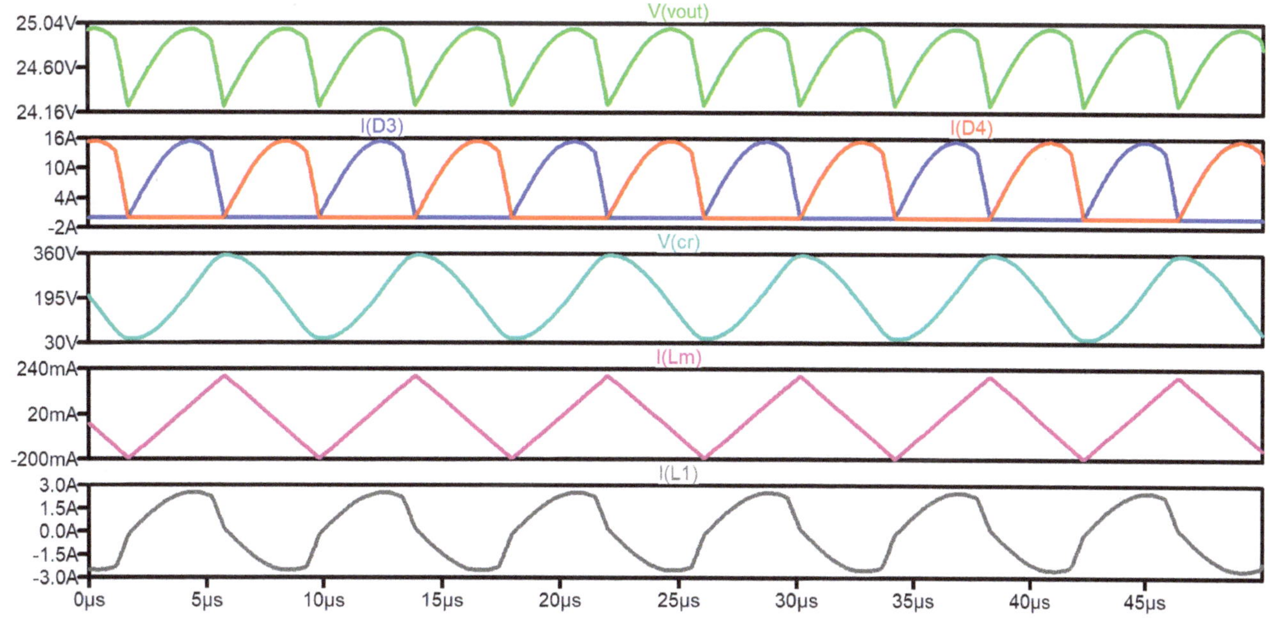

AC Response of the LLC Converter

THE ANALYTICAL STUDY of the control-to-output transfer function of the LLC converter, remains a difficult exercise as conventional averaging techniques cannot be applied to this resonant structure. Mathematical tools, like extended describing function (EDF), lead to complex expressions whose selection depends on the operating point. Fortunately, once the LLC components are selected and you confirmed that the open-loop converter delivers the correct operating variables, you can run the LTspice FRA to extract the needed transfer function. In this application circuit, the V_2 bias sets the switching frequency and the stimulus is classically injected in the VCO control pin.

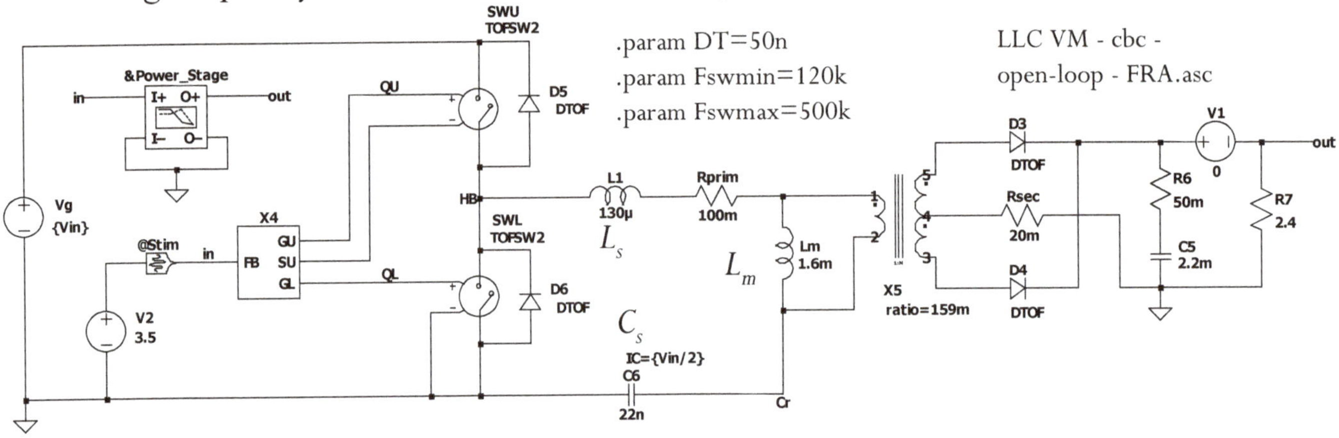

This design uses an unusually-high L_m/L_s ratio which leads to the frequency response shown below. A simple integrator would do, based on this first observation, but more operating points would obviously be necessary to confirm this choice. I will use a type 2 compensator with a PS parameter set to -45°, adding some margin. The stimulus amplitude is 200 mV. Note that the simulation time is quite long and the .RAW file was 9 Gb considering the wanted granularity but, the curves are smooth.

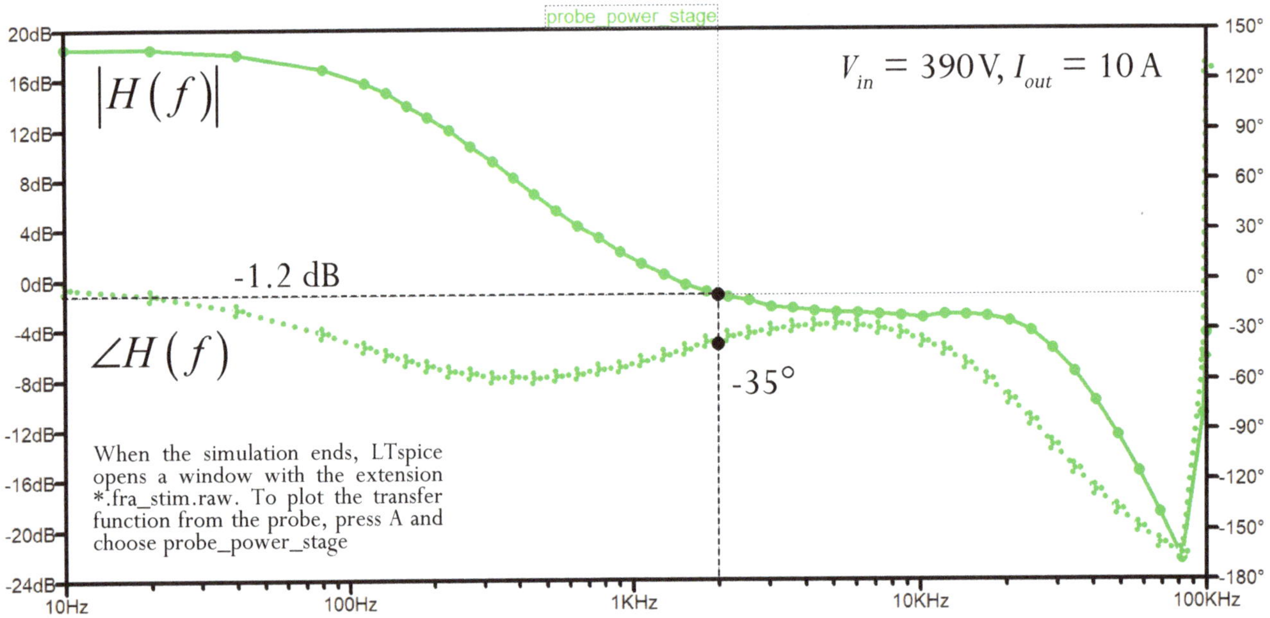

A Bootstrapped LLC Converter

THE BOOTSTRAP TECHNIQUE offers a simple and elegant way to drive the high-side transistor in a half-bridge converter. A capacitor is first charged from a ground-referenced V_{cc} source via a high-voltage diode, with the negative pin of this capacitor connected to the swinging half-bridge node. When the low-side switch is turned on, the HB node is grounded and the capacitor charges to V_{cc}- V_f, where V_f the diode forward drop. When this low-side switch turns off, the drain voltage rises fast and lifts the negative terminal of C_{boot} to the half-bridge potential, now supplying the upper-side driver with a *floating* voltage referenced to the upper transistor source. The IR2110 from International-Rectifier (now Infineon) pioneered this technique with a fully-isolated well, bringing robustness against negative biases on the HB node. Nowadays, bootstrap is still used and many half-bridge drivers are fully isolated to cope with the extremely-high dV/dt slopes generated by wide-bandgap semiconductors.

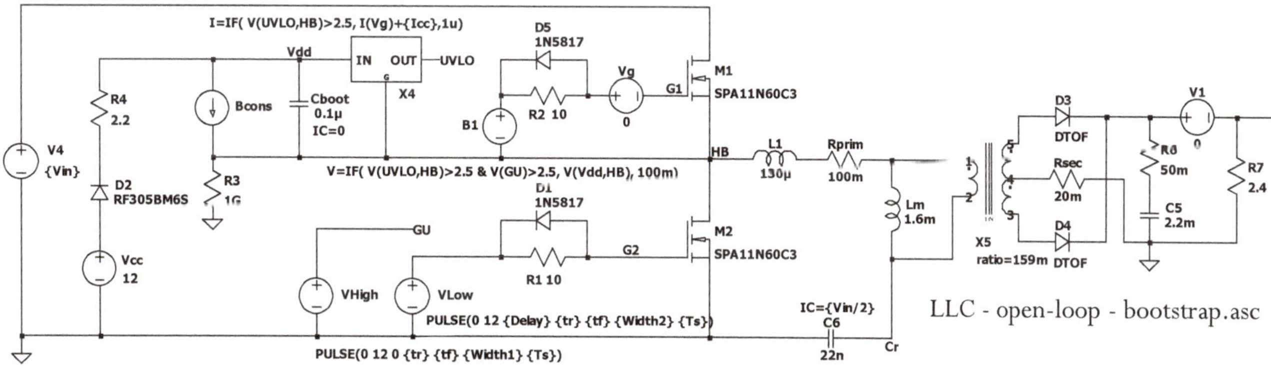

LLC - open-loop - bootstrap.asc

The output voltage is well regulated to 24 V, while the voltage across C_{boot} reaches ≈11.5 V. Resistance R_4 is necessary to reduce the noisy charging spike when M_2 turns on. In a real design, observe how negative the HB node swings and check the value versus the datasheet maximum ratings: negative substrate injection is often a source of trouble.

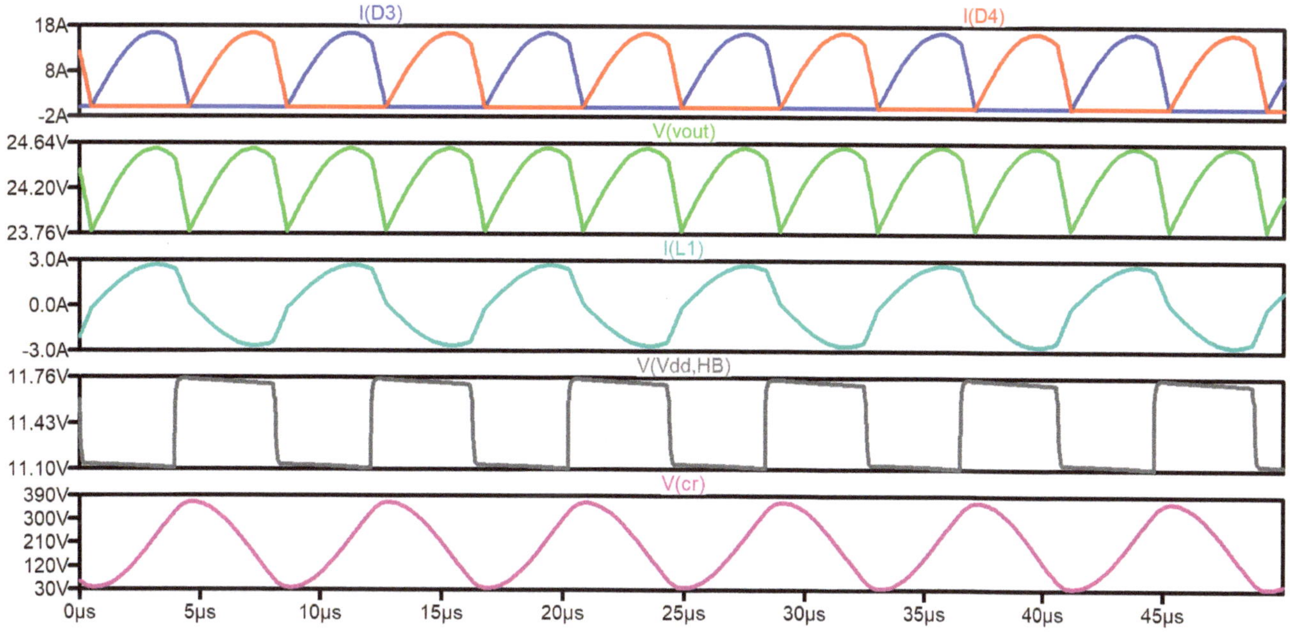

Transient Response of the LLC

NOW THAT WE HAVE tested the converter's core, we can close the loop and check the transient response. Closing the loop with direct frequency control (DFC) – it means the error voltage *directly* controls the VCO – can become a difficult exercise as the magnitude-phase response significantly varies with the operating point. For this reason, it is usually difficult to push the crossover frequency beyond a few kHz, especially if the input voltage varies. For instance, if you have a front-end PFC, the voltage will be stabilized at, perhaps, 400 V. However, during a line dropout event, you need to ensure stability with V_{in} dropping to 350 V, before the brown-out protection trips. Therefore, studying stability with this unusual low input voltage range is key to design a rugged and reliable product.

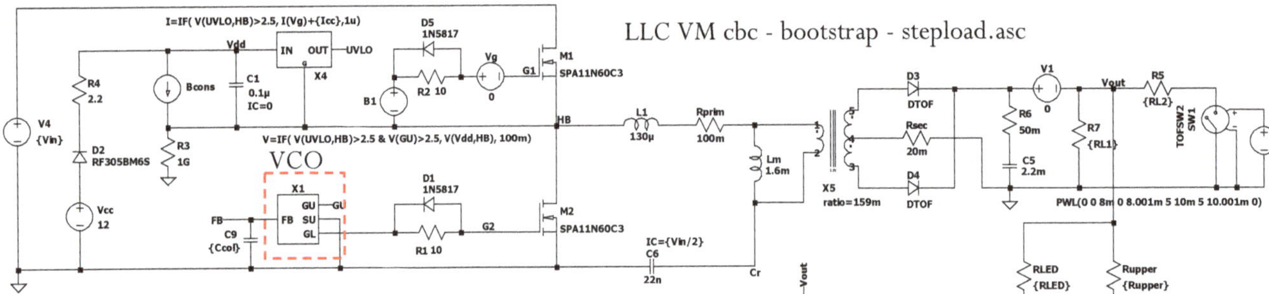

.param DT=120n
.param Fswmin=120k
.param Fswmax=500k
.param VDRV=12
.param Rpullup=20k

It is important to reflect the frequency limits from your converter for the given feedback dynamic as it sets the VCO small-signal gain.

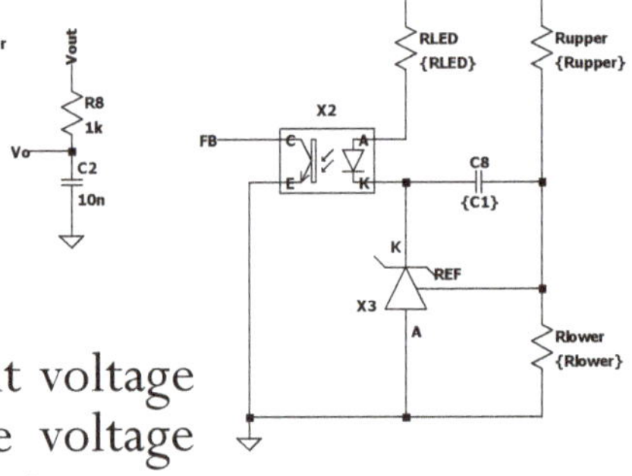

A simple type 2 is used in this loop.

The response to a load step is very stable at an input voltage of 400 V (also at 350 V). You need to check the voltage excursion on the resonant capacitor as well as its rms current. V_{out} is observed via the R_8-C_2 filter.

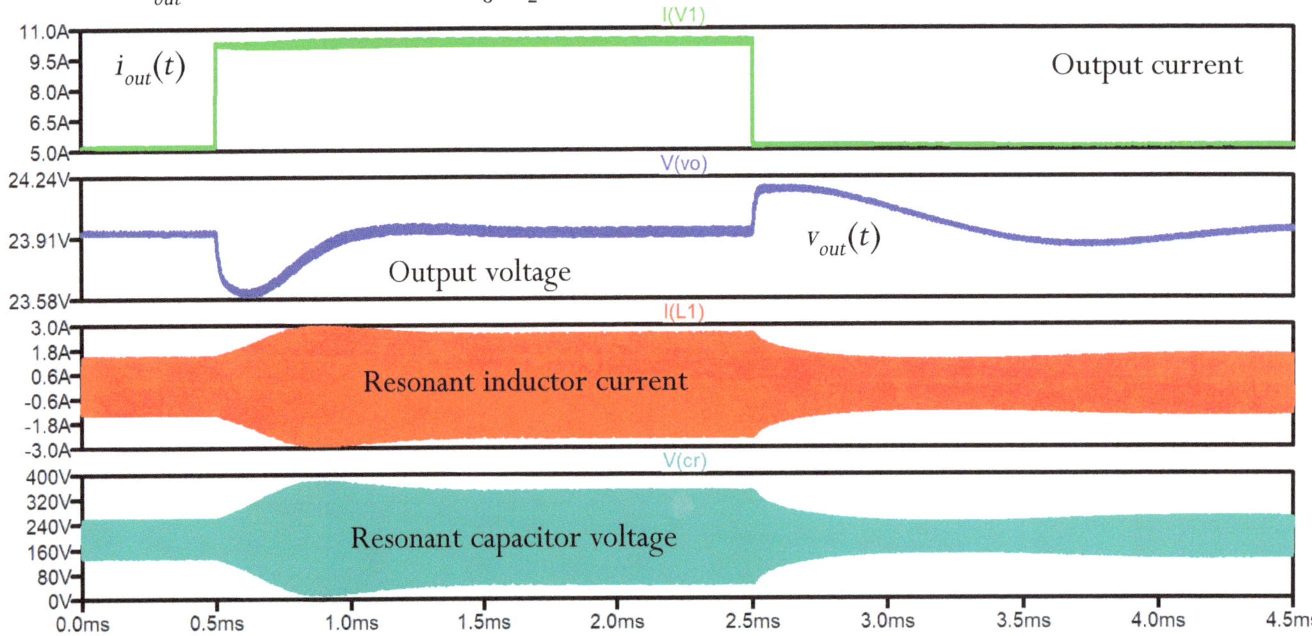

Bang-Bang Charge-Controlled LLC

THE BANG-BANG CHARGE-CONTROLLED LLC was described by researchers of Queens University (Kingston, CA) in a paper published in 2013 [8]. In this document, the authors show that, rather than controlling the switching frequency *directly* from the error voltage (direct frequency control or DFC), they could control the power flow by adjusting the peak and valley voltages of the resonant capacitor, *indirectly* controlling the switching frequency. This is now a first-order system, easier to control than with DFC. The subcircuit I have built, reflects this operating mode in a simplified way, and lets you test the operating point and a transient response.

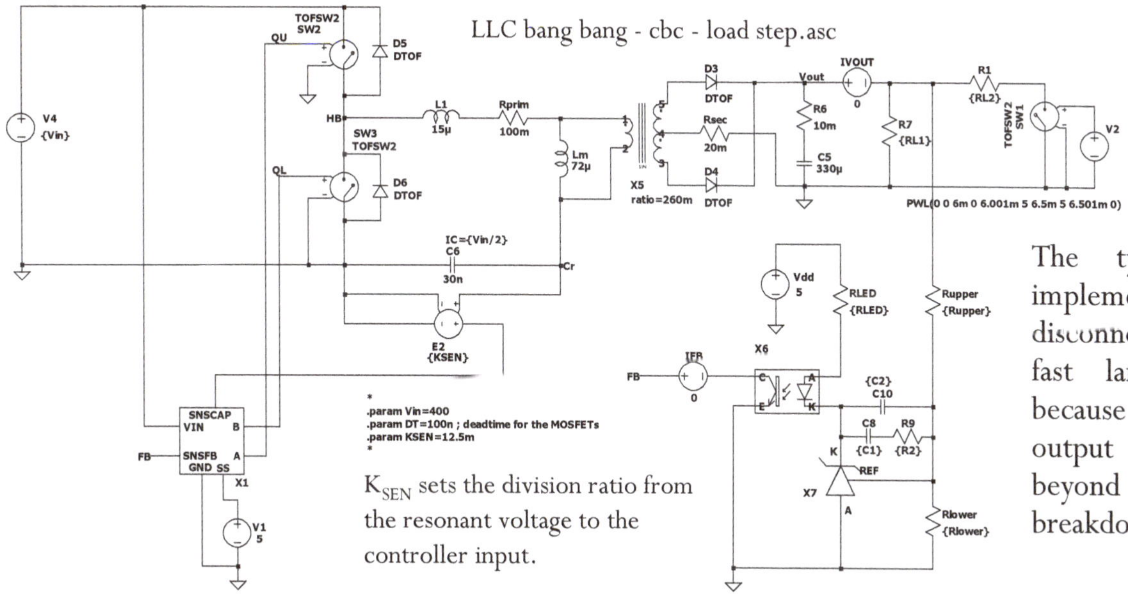

The type 2 is implemented by disconnecting the fast lane, mainly because of the 48-V output level, well beyond the TL431 breakdown voltage.

The transient response is very smooth on this 48-V/500-W converter, stepped from 50 to 100% load while supplied at 400 V. A theoretical crossover frequency of 10 kHz was set without issue to obtain the desired phase margin.

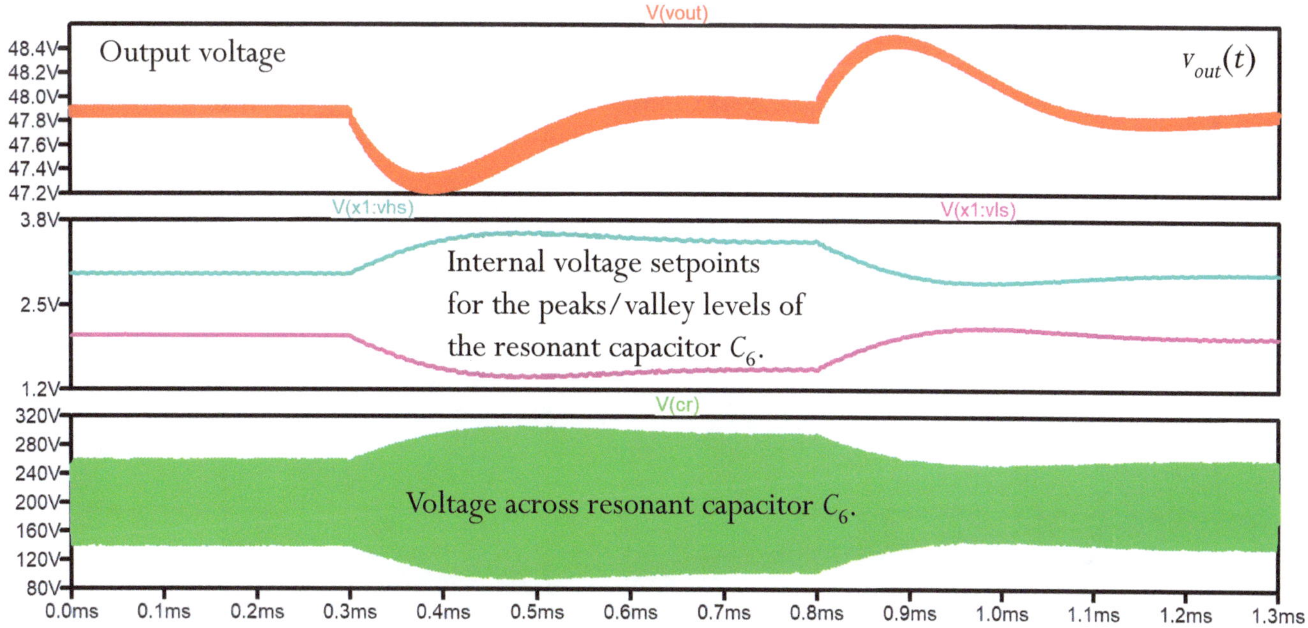

250-W LLC CC-CV e-Bike Charger

CURRENT-MODE LLC controllers offer a good way to build converters that are easier to control than their DFC counterparts. The example that I built, operates with two independent modes – constant voltage (CV) and constant current (CC) – well suited for charging the Li-Ion battery of an electrically-assisted bicycle.

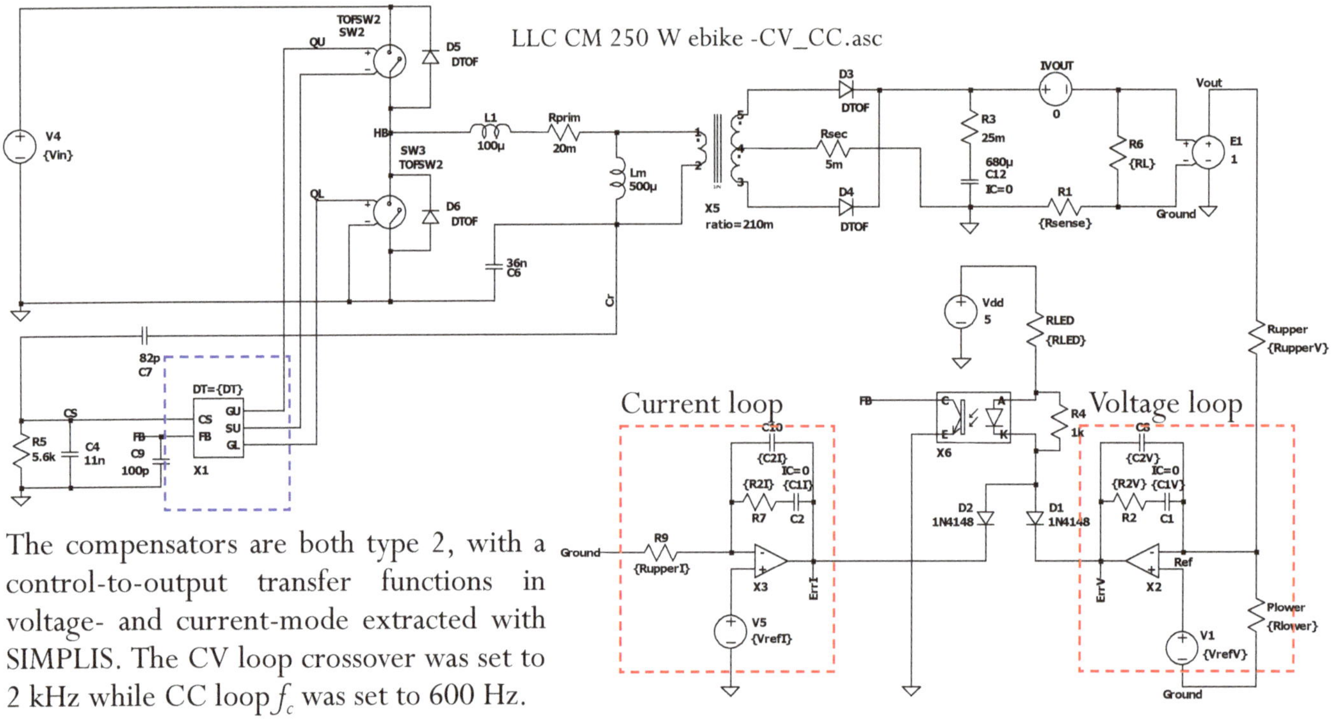

The compensators are both type 2, with a control-to-output transfer functions in voltage- and current-mode extracted with SIMPLIS. The CV loop crossover was set to 2 kHz while CC loop f_c was set to 600 Hz.

In the below graph, I have replaced the load (referenced to node *Ground*) by a 20-mF capacitor, pre-charged at 24 V. The current is well regulated at 7 A but V_{out} overshoots above the 36-V target considering the speed at which the voltage rises to no-load condition (50-ms simulation), which is obviously not that fast in practice. Also, modern parts include skip-cycle to prevent this situation, what I have not included here.

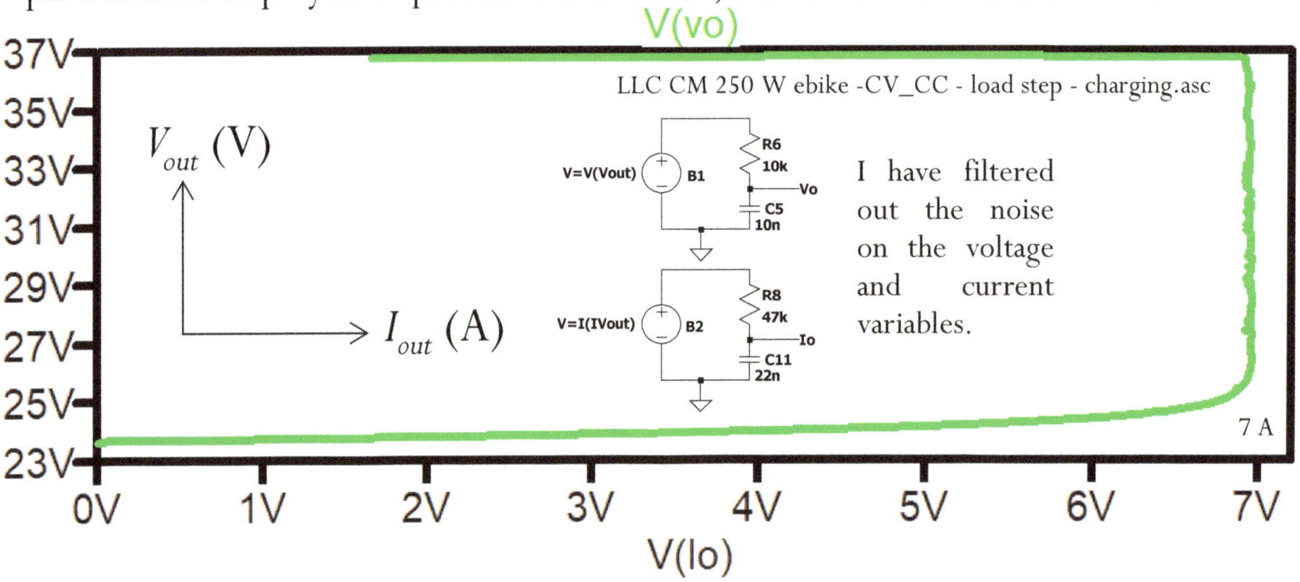

A Few Books to Consider

I HAVE GATHERED some of the books I believe you could consider for strengthening your knowledge in the field of power conversion. There are many sources but few are specifically focused on our field specifically and, most importantly, include practical details for immediate implementation.

An Intuitive Guide to Compensating Switching Power Supplies – Christophe Basso.

This practical manual on control loop design, starts with a bit of theory and documents many examples for stabilizing control systems.

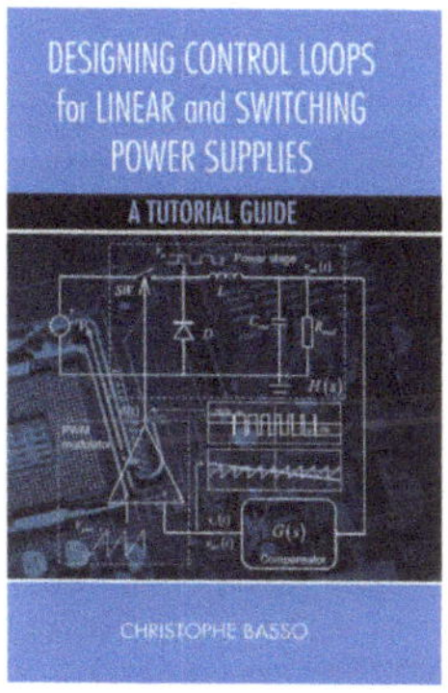

Designing Control Loops for Linear and Switching Power Supplies – Christophe Basso.

A good balance between theory you must know and design considerations with many examples.

The LTspice XVII Simulator Commands and Applications – Gilles Brocard.

This book details what you need to know to get the quintessence of this free simulator. Plenty of useful examples.

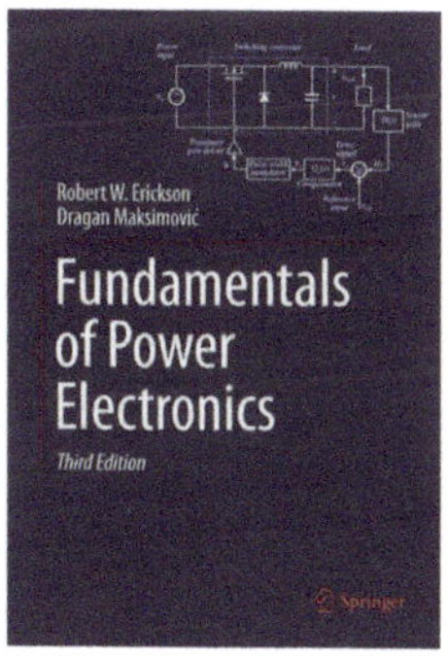

Fundamentals of Power Electronics – Robert Erickson and Dragan Maksimović.

This is an excellent and comprehensive text on power electronics. A must-read for engineers willing to understand theory.

Digital Control of High-Frequency Switched-Mode Power Converters – L. Corradini, D. Maksimović, P. Mattavelli, R. Zane

If you look for a rigorous and extensive coverage of digital control, it is an excellent read.

Transfer Functions of Switching Converters – Christophe Basso

If you wonder how to derive the transfer functions of switching converters, this is the book.

Papers and Articles

I HAVE LISTED papers I believe are relevant to the topic of switching converters simulations. The references in each paper link to another source of knowledge which let you progress in this field.

1. L. G. Meares, *New Simulation Techniques using SPICE*, IEEE Applied Power Electronics Conference, New Orleans, 1986

2. D. Adar, S. Ben-Yaakov, *Generic average modeling and simulation of discrete controllers*, APEC, Anaheim, 2001

3. A. Kamath, *How to Prevent Transformer Saturation in Push-Pull Converters*, Texas-Instruments, Analog Design Journal, 2021

4. A. Laprade, *Enhanced OTA Models Improve Design Of Feedback Compensation Networks*, How2Power, July 2024 issue.

5. A. Vorpérian, *Analysis of Current-Controlled Converters using the Model of the Current-Controlled PWM Switch,* Power Conversion and Intelligent Motion Conference, pp. 183-195, 1990

6. K. Yao et al., *Critical Bandwidth for the Load Transient Response of Voltage Regulator Modules*, IEEE Transactions on Power Electronics, Vol. 19, NO. 6, 2004

7. C. Hymowitz, *Monte Carlo gone Wrong*, on-line EDN Magazine, 2019

8. Z. Hu, Y. Liu, *Bang Bang Charge Control for LLC Resonant Converters*, IEEE Transactions on Power Electronics, Vol. 30, NO. 2, 2015

9. R. Nielsen, LLC and LCC Resonant Converters, http://www.runonielsen.dk, August 2013

10. D. Venable, *The k Factor; A new Mathematical Tool for Stability Analysis and Synthesis*, Proceeding of *Powercon 10*, San-Diego, CA, march 22-24,1983

11. C. Basso, *The TL431 in the Control of Switching Power Supplies*, *onsemi* Seminar, September 2009

12. L. Dixon, *Closing the Feedback Loop*, SEM300, Unitrode Seminars, SLUP068, Texas-Instruments

13. C. Basso, *Understanding Op Amp Dynamic Response In A Type-2 Compensator*, 2-part article, How2Power.com, January 2017

14. C. Basso, *Modern Control Methods For LLC Converters Simplify Compensator Design*, How2Power.com, December 2021

15. C. Basso, *Analysis, Simulation and Experimentation Enable Successful Design of Power Supply Compensation*, How2Power.com, July 2020

16. Y. Panov, M. Jovanović, L. Huber, *Principles of Converter Control*, Internal Course, Delta Electronics, 2002

17. M. Jovanović, *Introduction to Digital Control of Switch-Mode Power Converters*, Internal course, Delta Electronics, 2014

18. J. Turchi, *Compensation of a PFC Stage Driven by the NCP1654*, onsemi, AND8321/D, 2009

Internet Links

BELOW IS A SHORT LIST of links where you will find interesting materials to download or videos to watch.

My webpage in which you will find books, application notes, seminars and the free ready-made templates I used in the book: http://powersimtof.com/Spice.htm

How2Power is the on-line monthly newsletter edited by David Morrison, entirely dedicated to power electronics. A must-read for power engineers: http://www.how2power.com/

Electronic StackExchange is an interesting professional forum in which users ask questions on various design subjects, including power and control. The answers are often very well documented: https://electronics.stackexchange.com/questions

This page unveils many undocumented features from LTspice:
https://ltwiki.org/?title=Undocumented_LTspice#Dual_Value_Resistors_.28for_ac_analysis.29

The LTspice forum on groups.io is an excellent resource to ask questions or read answers from LTspice users: https://groups.io/g/LTspice

The ADIEngineersZone hosts a forum where ADI application engineers answer questions and share simulation tips: https://ez.analog.com/design-tools-and-calculators/ltspice

Plenty of YouTube channels to learn from!

Brian Douglas is an enthusiast of control systems and an excellent teacher too:

https://www.youtube.com/@BrianBDouglas/videos

Robert Bolanos is well-known in the field of power supply and posts many interesting tutorial videos on loop design:

https://www.youtube.com/@RobertBolanos

Ali Shirsavar of Biricha Digital covers many aspects of loop control, in the digital and analog domains: https://www.youtube.com/@BirichaLectures/videos

Katherine Kim teaches at the National Taiwan University and runs an instructive YouTube channel with plenty of good videos on power electronics:

https://www.youtube.com/user/katkimshow

Marcos Alonso teaches at the university of Oviedo, Spain, and posts many interesting videos on control, illustrated with LTspice® or QSPICE®:

https://www.youtube.com/@MarcosAlonsoElectronics/featured

Sam Ben-Yaakov offers many excellent videos on power electronics, ranging from simple to complicated structures, a must-visit site: https://www.youtube.com/user/sambenyaakov

Eddie Aho presents many interesting things on power circuits, instruments but also books he reviews: https://www.youtube.com/c/kissanalog

www.ingramcontent.com/pod-product-compliance
Lightning Source LLC
Chambersburg PA
CBHW060906250226
40239CB00031B/2096